Lecture Notes in Mathematics

Volume 2289

This series reports on new developments in all areas of mathematics and their applications - quickly, informally and at a high level. Mathematical texts analysing new developments in modelling and numerical simulation are welcome. The type of material considered for publication includes:

1. Research monographs
2. Lectures on a new field or presentations of a new angle in a classical field
3. Summer schools and intensive courses on topics of current research.

Texts which are out of print but still in demand may also be considered if they fall within these categories. The timeliness of a manuscript is sometimes more important than its form, which may be preliminary or tentative.

Titles from this series are indexed by Scopus, Web of Science, Mathematical Reviews, and zbMATH.

More information about this series at http://www.springer.com/series/304

Toshihiro Iwai

Geometry, Mechanics, and Control in Action for the Falling Cat

Springer

Toshihiro Iwai
Kyoto University
Kyoto, Japan

ISSN 0075-8434 ISSN 1617-9692 (electronic)
Lecture Notes in Mathematics
ISBN 978-981-16-0687-8 ISBN 978-981-16-0688-5 (eBook)
https://doi.org/10.1007/978-981-16-0688-5

This Springer imprint is published by the registered company Springer Nature Singapore Pte Ltd.
The registered company address is: 152 Beach Road, #21-01/04 Gateway East, Singapore 189721, Singapore

Preface

The falling cat and the falling apple of Newton share the same mechanical property, i.e., vanishing angular momentum. We begin by giving a brief review of the falling apple of Newton. An apple fell down at his feet, when Newton was sitting in an orchard on a summer afternoon. This fact is supposed to have inspired Newton to find the universal law of gravitation on the basis of Kepler's laws of planetary motion. For us, people of today, the falling of an apple is a two-body (an apple and the Earth) problem, which reduces to a one-body problem with respect to a mass-weighted difference vector between two bodies. As the initial velocity of the present difference vector vanishes, the angular momentum vanishes as well. Then, the conservation law of the angular momentum provides the equation

$$\boldsymbol{r} \times \frac{d\boldsymbol{r}}{dt} = 0, \tag{1}$$

where $\boldsymbol{r}$ denotes the mass-weighted difference vector. This implies that $\boldsymbol{r}$ and $d\boldsymbol{r}/dt$ are in parallel, so that there exists a scalar function λ such that $d\boldsymbol{r}/dt = \lambda\boldsymbol{r}$. This means that the motion of the apple takes place along the line segment joining the respective centers of mass of the Earth and the apple. In contrast to this, the moon does not fall to Earth. This is because the angular momentum of the difference vector between the moon and the Earth does not vanish. In order to find the trajectories of the apple and the moon in respective cases, we need to solve Newton's equations of motion with gravitational force.

One of Kepler's laws refers to the non-vanishing of the angular momentum of planets. The first law says that the orbit of every planet is an ellipse with the Sun at one of the two foci, and the second law says that a line segment joining a planet and the Sun sweeps out equal areas during equal intervals of time. These laws imply that the angular momentum is conserved and the force is central. The planet in question (or the moon) has a non-vanishing angular momentum. Further, the central force is shown to be inversely proportional to the square of the distance. In order to show that the proportional constant is a universal constant, the third law is needed, which says that the square of the orbital period of a planet is proportional to the cube of the semi-major axis of its orbit (see Appendix 5.1 for details).

In contrast to the falling apple of Newton, when interest attaches to the falling cat, the gravitational force may be put aside. We then reconsider the condition of the vanishing angular momentum in view of the fact that the angular momentum of the cat surely vanishes in the air. The angular momentum of an N-particle system with $N \geq 3$ can bring about a complicated motion of each particle. For an N-particle system, there are $N - 1$ mass-weighted vectors (Jacobi vectors to be introduced in the text), in terms of which Eq. (1) is extended to

$$\sum_{i=1}^{N-1} \boldsymbol{r}_i \times \frac{d\boldsymbol{r}_i}{dt} = 0. \tag{2}$$

An initial question we are interested in concerns a property exhibited by a motion satisfying (2). Physically speaking, any motion satisfying this equation is viewed as vibrational motion, as the total angular momentum of the particle system vanishes. In contrast to solutions to (1), solutions to (2) exhibit interesting motions with remarkable geometric properties and are used to explain a reason why vibrations can give rise to effective rotations. In addition, a further question arises as to whether the following partial differential equations may have solutions or not:

$$\sum_{i=0}^{N-1} \boldsymbol{r}_i \times d\boldsymbol{r}_i = 0. \tag{3}$$

In other words, the question is stated as follows: Are there $(3N - 6)$-dimensional surfaces satisfying the above equations in $\mathbb{R}^{3N-3}$? This equation has been of central interest in the study on the possibility of the separation of vibration from rotation. However, Eq. (3) is not integrable. The quantity associated with the non-integrability has a geometric meaning. Since the falling cat makes vibrational motions only in the air, a solution to the falling cat problem must involve questions regarding Eqs. (2) and (3) without reference to gravitational force.

As is well known, cats always can land on their feet when launched in the air. If cats are not given a non-vanishing angular moment at an initial instant, they cannot rotate during their motion, but the motion they can make in the air is vibration only. However, cats accomplish a turn after a vibrational motion, when landing on their feet. As is alluded to above, in order to solve this apparent mystery, one needs to gain a strict understanding of rotations and vibrations. The connection theory in differential geometry can provide rigorous definitions of rotation and vibration for many particle systems [22] and shows that vibrational motions can result in rotations, without performing rotational motions. The deformable bodies of cats are not easy to describe or to analyze in terms of mechanics. A feasible way to approach the question about the falling cat is to start with many particle systems and then proceed to systems of rigid bodies, and further to jointed rigid bodies, which can approximate the body of a cat. Since the present model of the falling cat is a mechanical object, mechanics of many-body systems and of jointed rigid bodies need to be set up. In order to take into account the fact that cats can contort their bodies, torque inputs should be applied as control inputs, which are to be suitably designed. In this book, the port-controlled Hamiltonian method will be adopted for the jointed rigid bodies to perform a turn and to halt the motion at the instance of landing. A brief review of control systems will be given through simple examples to explain the role of control inputs.

This book consists of four chapters, the headings of which refer, in accordance with the title of this book, to Geometry, Mechanics, Control, and the Falling Cat, respectively. In the first chapter, geometries of many-body systems are set up, starting with planar many-body systems. As planar many-body systems have rather simple geometric structures, they will serve as an informative guide to the geometry of spatial many-body systems. After a review of the rotation group $SO(3)$, a geometric setting is provided for spatial many-body systems. In this chapter, the main interest centers on the non-separability of rotation and vibration. The second chapter deals with mechanics of many-body systems on the basis of the geometric setting. Rigid bodies are treated as special cases of many-body systems, and rigid body mechanics is reformulated on the variational principle in the Lagrangian and Hamiltonian formalisms. Subsequently, Lagrangian and Hamiltonian mechanics for spatial many-body systems are set up on the variational principle as well. The third chapter contains examples of mechanical control systems with interest in the understanding of the design of control inputs from mechanical point of view. The last chapter is devoted to the falling cat problem with two jointed cylinders as a model system. Geometry, mechanics, and control set up in the preceding chapters are employed in the analysis of the falling cat. Advanced material concerning many-body systems and related topics together with Newton's law of gravitation are discussed and reviewed in the appendices.

A short remark on notations used in this book is to be made in advance of the text. The space of real $n \times m$ matrices is denoted by $\mathbb{R}^{n\times m}$, and the transpose of a matrix A by A^T.

Before beginning with the text, the author would like to stress that the present book gives a theory for the falling cat and does not recommend readers to make experiments on cats that test the bonds of trust between cats and human beings.

The author would like to thank his old students and colleagues who were interested in the falling cat problem for discussions with, questions from, and suggestions by them. This book is a result of those activities. In particular, thanks go to Mr. Matsunaka. The graphs in Figs. 4.3, 4.4, 4.5 and in Figs. 4.6, 4.7, 4.8 and the snapshots in Fig. 4.9 given in Chap. 4 are newly produced for the present book by Mr. Matsunaka, who was a coauthor of an old joint paper on the falling cat [39].

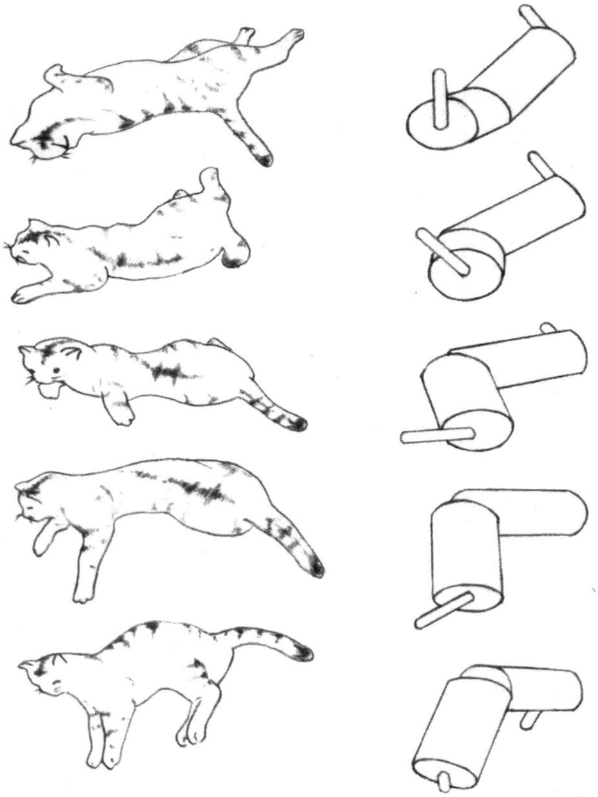

Fig. 1 A sketch of the falling cat and a model cat using jointed cylinders. Small rods attached to the cylinders are indicators to the attitude

The author is indebted to his wife and daughter for Fig. 1. Figure 1 is a sketch of the idea for the falling cat and Fig. 4.9 shows a realization of the idea developed in this book.

Kyoto, Japan

Toshihiro Iwai

Contents

Chapter 1
Geometry of Many-Body Systems

This chapter deals with the geometric setting for planar and spatial many-body systems on the basis of connection theory. Rather, the contents of this chapter may be of practical help in understanding the connection theory.

1.1 Planar Many-Body Systems

Suppose we are given a system of N point particles in the plane $\mathbb{R}^2$. Let $\boldsymbol{x}_\alpha$ and $m_\alpha > 0$, $\alpha = 1, 2, \ldots, N$, denote the position and the mass of each particle. Then, the configuration of these particles is described as $(\boldsymbol{x}_1, \boldsymbol{x}_2, \cdots, \boldsymbol{x}_N)$ (Fig. 1.1). The totality of the configurations is called the configuration space, which we denote by

$$X = \{x;\ x = (\boldsymbol{x}_1, \boldsymbol{x}_2, \cdots, \boldsymbol{x}_N),\ \boldsymbol{x}_\alpha \in \mathbb{R}^2\}. \tag{1.1}$$

The X is a linear space of dimension $2n$, in which addition and scalar multiplication are performed componentwise:

$$\begin{aligned}(\boldsymbol{x}_1 + \boldsymbol{y}_1, \boldsymbol{x}_2 + \boldsymbol{y}_2, \cdots, \boldsymbol{x}_N + \boldsymbol{y}_N) &= (\boldsymbol{x}_1, \boldsymbol{x}_2, \cdots, \boldsymbol{x}_N) + (\boldsymbol{y}_1, \boldsymbol{y}_2, \cdots, \boldsymbol{y}_N),\\ \lambda(\boldsymbol{x}_1, \boldsymbol{x}_2, \cdots, \boldsymbol{x}_N) &= (\lambda\boldsymbol{x}_1, \lambda\boldsymbol{x}_2, \cdots, \lambda\boldsymbol{x}_N).\end{aligned} \tag{1.2}$$

We can view this system as a vector space of $2 \times N$ matrices consisting of N column vectors $\boldsymbol{x}_\alpha \in \mathbb{R}^2$. Let us denote the canonical basis vectors of $\mathbb{R}^2$ by

$$\boldsymbol{e}_1 = \begin{pmatrix}1\\0\end{pmatrix}, \qquad \boldsymbol{e}_2 = \begin{pmatrix}0\\1\end{pmatrix}.$$

T. Iwai, *Geometry, Mechanics, and Control in Action for the Falling Cat*,
Lecture Notes in Mathematics 2289, https://doi.org/10.1007/978-981-16-0688-5_1

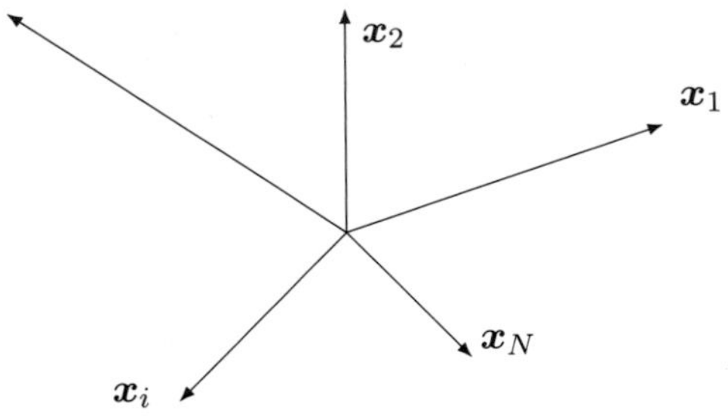

Fig. 1.1 Position vectors

Then, a basis of X is given, for example, by

$$(\boldsymbol{e}_1, 0, \cdots, 0), \quad (\boldsymbol{e}_2, 0, \cdots, 0), \quad \cdots, \quad (0, \cdots, \boldsymbol{e}_2). \tag{1.3}$$

The X is endowed with the mass-weighted inner product (or inner product, for short) defined to be

$$K(x, y) = \sum_{\alpha=1}^{N} m_\alpha \boldsymbol{x}_\alpha \cdot \boldsymbol{y}_\alpha, \quad x, y \in X, \tag{1.4}$$

where $\boldsymbol{x} \cdot \boldsymbol{y}$ denotes the standard inner product on $\mathbb{R}^2$. As is easily verified, the $K(x, y)$ has the properties

$$K(x, y) = K(y, x), \tag{1.5a}$$

$$K(\lambda x + \mu y, z) = \lambda K(x, z) + \mu K(y, z), \qquad \lambda, \mu \in \mathbb{R}, \tag{1.5b}$$

$$K(x, x) \geq 0, \qquad K(x, x) = 0 \iff x = 0. \tag{1.5c}$$

The center-of-mass system is the linear subspace of X which is defined to be

$$Q = \{x \in X; \sum_{\alpha=1}^{N} m_\alpha \boldsymbol{x}_\alpha = 0\}. \tag{1.6}$$

In order to characterize the Q in X, it is of help to introduce the center-of-mass vector for an arbitrary configuration $y = (\boldsymbol{y}_1, \cdots, \boldsymbol{y}_N) \in X$ through

$$\boldsymbol{b} = \left(\sum_{\beta=1}^{N} m_\beta \right)^{-1} \sum_{\alpha=1}^{N} m_\alpha \boldsymbol{y}_\alpha. \tag{1.7}$$

Then, by setting $\boldsymbol{x}_\alpha = \boldsymbol{y}_\alpha - \boldsymbol{b}$, one has the decomposition of y,

$$(\boldsymbol{y}_1, \boldsymbol{y}_2, \cdots, \boldsymbol{y}_N) = (\boldsymbol{x}_1, \boldsymbol{x}_2, \cdots, \boldsymbol{x}_N) + (\boldsymbol{b}, \boldsymbol{b}, \cdots, \boldsymbol{b}). \tag{1.8}$$

We can easily verify that $x = (\boldsymbol{x}_1, \boldsymbol{x}_2, \cdots, \boldsymbol{x}_N) \in Q$, and further that

$$(\boldsymbol{x}_1, \boldsymbol{x}_2, \cdots, \boldsymbol{x}_N) \perp (\boldsymbol{b}, \boldsymbol{b}, \cdots, \boldsymbol{b}),$$

where the orthogonality is defined with respect to the inner product K. On setting $b = (\boldsymbol{b}, \cdots, \boldsymbol{b})$, Eq. (1.8) is written as $y = x + b$, which is an orthogonal decomposition of $y \in X$. Let us denote the orthogonal complement of Q by

$$Q^{\perp} := \{y \in X;\ K(y, x) = 0,\ x \in Q\}. \tag{1.9}$$

Then, we have the orthogonal decomposition of X,

$$X = Q \oplus Q^{\perp}, \tag{1.10}$$

to which $y = x + b$ is subject. It is easily shown that

$$Q^{\perp} = \{x \in X |\ x = (\boldsymbol{c}, \boldsymbol{c}, \cdots, \boldsymbol{c}),\ \boldsymbol{c} \in \mathbb{R}^2\}. \tag{1.11}$$

The basis (1.3) is not adequate to the decomposition (1.10). In fact, the basis belongs to neither Q nor $Q^{\perp}$. We wish to find a basis adapted to (1.10). An orthonormal basis of $Q^{\perp}$ is easy to find on account of (1.11). In order to find a basis of Q, we can start, for example, with $(-m_2\boldsymbol{e}_1, m_1\boldsymbol{e}_1, 0, \cdots, 0)$, $(-m_2\boldsymbol{e}_2, m_1\boldsymbol{e}_2, 0, \cdots, 0)$, which clearly belong to Q, and apply the Schmidt method. Performing a similar procedure for other vectors in Q, we can find an orthonormal basis. Thus, we obtain the orthonormal system as follows:

Proposition 1.1.1 *The configuration space X for the planar N-body system admits the following orthonormal system of basis vectors:*

$$\begin{aligned}
c_1 &= N_0(\boldsymbol{e}_1, \boldsymbol{e}_1, \cdots, \boldsymbol{e}_1), \\
c_2 &= N_0(\boldsymbol{e}_2, \boldsymbol{e}_2, \cdots, \boldsymbol{e}_2), \\
f_{2j-1} &= N_j(\overbrace{-m_{j+1}\boldsymbol{e}_1, \cdots, -m_{j+1}\boldsymbol{e}_1}^{j \text{ terms}}, (\sum_{\alpha=1}^{j} m_\alpha)\boldsymbol{e}_1, 0, \cdots, 0), \\
f_{2j} &= N_j(\overbrace{-m_{j+1}\boldsymbol{e}_2, \cdots, -m_{j+1}\boldsymbol{e}_2}^{j \text{ terms}}, (\sum_{\alpha=1}^{j} m_\alpha)\boldsymbol{e}_2, 0, \cdots, 0),
\end{aligned} \tag{1.12}$$

where $j = 1, 2, \ldots, N-1$, and where

$$N_0 = \left(\sum_{\alpha=1}^{n} m_\alpha\right)^{-1/2}, \quad N_j = \left(m_{j+1}\left(\sum_{\alpha=1}^{j} m_\alpha\right)\left(\sum_{\alpha=1}^{j+1} m_\alpha\right)\right)^{-1/2}.$$

It is straightforward to verify that

$$K(c_a, c_b) = \delta_{ab}, \quad K(c_a, f_k) = 0, \quad K(f_k, f_\ell) = \delta_{k\ell}, \\ a, b = 1, 2, \ k, \ell = 1, 2, \dots, 2(N-1). \tag{1.13}$$

This means that

$$Q^\perp \cong \mathbb{R}^2, \qquad Q \cong \mathbb{R}^{2(N-1)}, \tag{1.14}$$

where $Q^\perp$ and Q have basis vectors $c_a, a = 1, 2$, and $f_k, k = 1, 2, \dots, 2(N-1)$, respectively. For any configuration $y = x + b \in Q \oplus Q^\perp$, the components of $x \in Q$ and $b \in Q^\perp$ with respect to these basis vectors are determined by

$$p_a = K(y, c_a) = K(b, c_a), \qquad a = 1, 2, \\ q_k = K(y, f_k) = K(x, f_k), \qquad k = 1, 2, \dots, 2(N-1), \tag{1.15}$$

and (p_a, q_k) serve as the Cartesian coordinates of $X = Q^\perp \oplus Q$.

In what follows, we take up Q endowed with the Cartesian coordinates (q_k). Any $x \in Q$ is expressed as $x = \sum_{k=1}^{2(N-1)} q_k f_k$. We now show that the coordinates (q_k) determine the Jacobi vectors on the plane $\mathbb{R}^2$. In fact, the vectors $\boldsymbol{r}_j$ defined by $\boldsymbol{r}_j := q_{2j-1}\boldsymbol{e}_1 + q_{2j}\boldsymbol{e}_2$, $j = 1, 2, \dots, N-1$, are written out as

$$\begin{aligned} \boldsymbol{r}_j &:= q_{2j-1}\boldsymbol{e}_1 + q_{2j}\boldsymbol{e}_2 = K(x, f_{2j-1})\boldsymbol{e}_1 + K(x, f_{2j})\boldsymbol{e}_2 \\ &= \Big(m_{j+1}\sum_{\alpha=1}^{j} m_\alpha\Big)^{1/2}\Big(\sum_{\alpha=1}^{j+1} m_\alpha\Big)^{-1/2}\Big(\boldsymbol{x}_{j+1} - \Big(\sum_{\alpha=1}^{j} m_\alpha\Big)^{-1}\sum_{\alpha=1}^{j} m_\alpha \boldsymbol{x}_\alpha\Big). \end{aligned} \tag{1.16}$$

The present expression of $\boldsymbol{r}_j$ shows that $\boldsymbol{r}_j$ is the vector with the initial point at the center-of-mass of the set of the particles at $\boldsymbol{x}_\alpha$, $\alpha = 1, \cdots, j$, to the end point at the position of the particle $\boldsymbol{x}_{j+1}$, within a constant multiple. The vectors $\boldsymbol{r}_j$ are exactly the Jacobi vectors (Fig. 1.2). Conversely, any of $\boldsymbol{x}_1, \cdots, \boldsymbol{x}_{N-1}$ can be expressed as a linear combination of $\boldsymbol{r}_1, \cdots, \boldsymbol{r}_{N-1}$, by using (1.16), and hence any of $\boldsymbol{x}_1, \cdots, \boldsymbol{x}_{N-1}, \boldsymbol{x}_N$ can be expressed as a linear combination of $(\boldsymbol{r}_1, \cdots, \boldsymbol{r}_{N-1})$ on account of $\boldsymbol{x}_N = -\frac{1}{m_N}\sum_{k=1}^{N-1} m_k \boldsymbol{x}_k$. It then turns out that the center-of-mass system Q is viewed as the totality of the Jacobi vectors $(\boldsymbol{r}_1, \cdots, \boldsymbol{r}_{N-1})$.

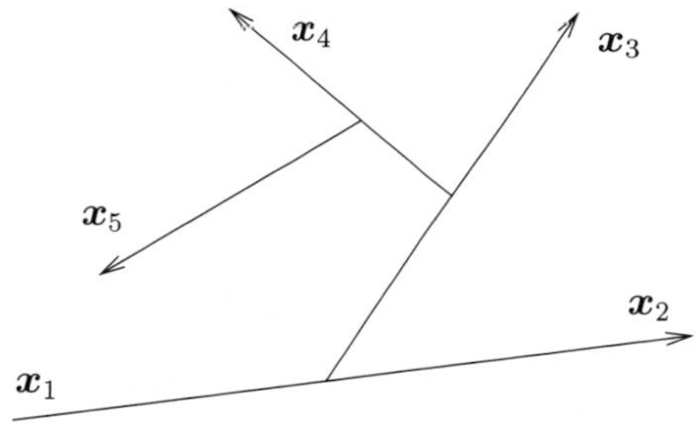

Fig. 1.2 Planar Jacobi vectors

In particular, for $N = 3$, the Jacobi vectors are described as

$$\begin{aligned} \boldsymbol{r}_1 &= q_1\boldsymbol{e}_1 + q_2\boldsymbol{e}_2 = \sqrt{\frac{m_1m_2}{m_1+m_2}}(\boldsymbol{x}_2 - \boldsymbol{x}_1), \\ \boldsymbol{r}_2 &= q_3\boldsymbol{e}_1 + q_4\boldsymbol{e}_2 = \sqrt{\frac{m_3(m_1+m_2)}{m_1+m_2+m_3}}\Big(\boldsymbol{x}_3 - \frac{m_1\boldsymbol{x}_1 + m_2\boldsymbol{x}_2}{m_1+m_2}\Big). \end{aligned} \tag{1.17}$$

Conversely, the position vectors are expressed as

$$\begin{aligned} \boldsymbol{x}_1 &= \frac{-1}{\sqrt{m_1+m_2}}\Big(\sqrt{\frac{m_2}{m_1}}\boldsymbol{r}_1 + \sqrt{\frac{m_3}{m_1+m_2+m_3}}\boldsymbol{r}_2\Big), \\ \boldsymbol{x}_2 &= \frac{1}{\sqrt{m_1+m_2}}\Big(\sqrt{\frac{m_1}{m_2}}\boldsymbol{r}_1 - \sqrt{\frac{m_3}{m_1+m_2+m_3}}\boldsymbol{r}_2\Big), \\ \boldsymbol{x}_3 &= -\frac{1}{m_3}(m_1\boldsymbol{x}_1 + m_2\boldsymbol{x}_2). \end{aligned} \tag{1.18}$$

In the center-of-mass system, any configuration coming from $x \in Q$ by a rotation has the same shape as the initial one x. We do not have to distinguish these configurations. This is because if we consider the configuration as a molecule with each particle being an atom, then the molecule property is independent of its position in $\mathbb{R}^2$, but depends on its shape only. In Euclidean geometry, congruence of triangles is discussed in a similar manner. Two triangles which are translated to one another by a translation and a rotation or a reflection are called congruent, which have the same shape.

We start with the definition of the $SO(2)$ action on the center-of-mass system Q. A reason why we do not treat the $O(2)$ action is as follows: If we view the N-body system as a molecule, the reflected N-body system would have a different chemical property from the initial one. Now we define the $SO(2)$ action by

$$x = (\boldsymbol{x}_1, \boldsymbol{x}_2, \cdots, \boldsymbol{x}_N) \longmapsto gx = (g\boldsymbol{x}_1, g\boldsymbol{x}_2, \cdots, g\boldsymbol{x}_N), \quad x \in Q,\ g \in SO(2), \tag{1.19}$$

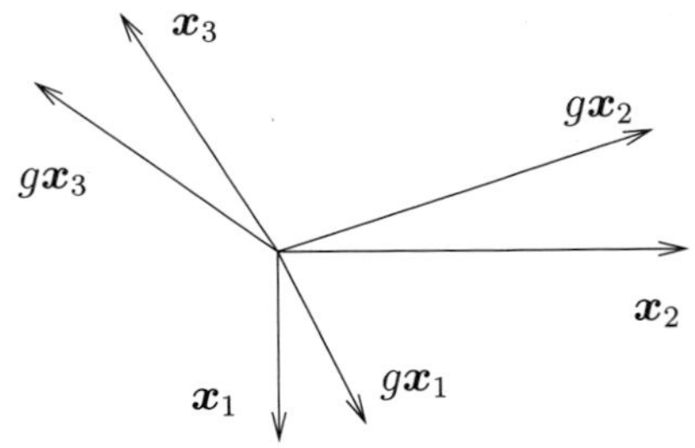

Fig. 1.3 $SO(2)$ action on a configuration

where

$$g = g(t) = \begin{pmatrix} \cos t & -\sin t \\ \sin t & \cos t \end{pmatrix}.$$

The $SO(2)$ action is well-defined on Q (Fig. 1.3). In fact, if $\sum m_\alpha \boldsymbol{x}_\alpha = 0$, then one has $\sum m_\alpha g \boldsymbol{x}_\alpha = 0$. The present $SO(2)$ action is a linear transformation of Q, the representation of which is determined by

$$g f_k = \sum_{j=1}^{2(N-1)} a_{jk} f_j, \quad k = 1, 2, \dots, 2(N-1),$$

where $f_k, k = 1, 2, \dots, 2(N-1)$, are the basis vectors already introduced in (1.12). Since $a_{jk} = K(f_j, g f_k)$, we can evaluate a_{jk} by using the definitions of the inner product K and of the basis vectors f_k. After a calculation, we find that

$$\begin{aligned} g(t) f_{2j-1} &= f_{2j-1} \cos t + f_{2j} \sin t, \\ g(t) f_{2j} &= -f_{2j-1} \sin t + f_{2j} \cos t. \end{aligned} \tag{1.20}$$

In each linear subspace spanned by f_{2j-1}, f_{2j}, the $SO(2)$ action is described, in terms of the Cartesian coordinates (q_{2j-1}, q_{2j}), as

$$\begin{pmatrix} q_{2j-1} \\ q_{2j} \end{pmatrix} \longmapsto \begin{pmatrix} \cos t & -\sin t \\ \sin t & \cos t \end{pmatrix} \begin{pmatrix} q_{2j-1} \\ q_{2j} \end{pmatrix}. \tag{1.21}$$

Consequently, the matrix representation of the $SO(2)$ is given by

$$(a_{jk}) = \begin{pmatrix} g(t) & & & \\ & g(t) & & \\ & & \ddots & \\ & & & g(t) \end{pmatrix},$$

which is a block diagonal matrix of size $2(N-1) \times 2(N-1)$.

The present $SO(2)$ action determines an equivalence relation on Q. For $x, y \in Q$, if there exists $g \in SO(2)$ such that $y = gx$, then x and y are called congruent and denoted by $x \sim y$. Then, it is easy to verify that

$$\text{(i) } x \sim x, \quad \text{(ii) } x \sim y \Longrightarrow y \sim x, \quad \text{(iii) } x \sim y,\ y \sim z \Longrightarrow x \sim z.$$

This equivalence relation determines the factor space consisting of equivalence classes. Geometrically speaking, the present factor space is the space of shapes of molecules. We are interested in non-trivial shapes. Put anther way, if all the particles collect at the center-of-mass, its configuration is $x = 0$, which is considered as shapeless and we eliminate it from Q. Let $\dot{Q}$ denote Q without $x = 0$,

$$\dot{Q} = \{x \in Q;\ x \neq 0\}. \tag{1.22}$$

Then, the $SO(2)$ action on $\dot{Q}$ is free, i.e., if $gx = x$ for $x \in \dot{Q}$ then $g = e$, the identity of $SO(2)$, as is easily verified. Thus, the factor space $\dot{Q}/SO(2)$ proves to be a manifold, which we call a shape space. We denote the shape space by $\dot{M}$ and the natural projection by π:

$$\pi : \dot{Q} \longrightarrow \dot{M} := \dot{Q}/SO(2). \tag{1.23}$$

In other words, if we denote the equivalence class of $x \in \dot{Q}$ by $[x]$, the projection is defined to be $\pi(x) = [x] \in \dot{Q}/SO(2)$. By definition, $[x] = [x']$ if and only if there exists $g \in SO(2)$ such that $x' = gx$.

The center-of-mass system Q has a rather simple structure with respect to the $SO(2)$ action. In order to give a compact description, we introduce the complex vector space structure into $Q \cong \mathbb{R}^{2(N-1)}$ by setting

$$z_j = q_{2j-1} + iq_{2j}, \quad j = 1, 2, \ldots, N-1, \quad i = \sqrt{-1}.$$

Then, we have the isomorphism $Q \cong \mathbb{C}^{N-1}$. Accordingly, the $SO(2)$ action on Q is represented as a $U(1)$ action,

$$z = (z_1, z_2, \cdots, z_{N-1}) \longmapsto e^{it}z = (e^{it}z_1, e^{it}z_2, \cdots, e^{it}z_{N-1}). \tag{1.24}$$

This is because the transformation (1.21) is compactly represented as

$$e^{it}z_j = (\cos t + i \sin t)(q_{2j-1} + iq_{2j}).$$

Now it is clear that $\dot{Q} \cong \mathbb{C}^{N-1} - \{0\}$, which we denote by $\dot{\mathbb{C}}^{N-1}$. The $U(1)$ action on $\dot{Q} \cong \dot{\mathbb{C}}^{N-1}$ is free, of course. In fact, if $e^{it}z = z$ for $z \in \dot{\mathbb{C}}^{N-1}$, then $e^{it} = 1$.

We define a map $\dot{\mathbb{C}}^{N-1} \to \mathbb{R}_+ \times \mathbb{C}^{N-1}$ to be

$$(z_1, \cdots, z_{N-1}) \longmapsto \Big(|z|, \frac{z_1}{|z|}, \cdots, \frac{z_{N-1}}{|z|}\Big), \qquad |z|^2 = \sum_{j=1}^{N-1} |z_j|^2, \tag{1.25}$$

where $\mathbb{R}_+$ denotes the set of positive real numbers. Then, we verify that

$$\dot{Q} \cong \dot{\mathbb{C}}^{N-1} \cong \mathbb{R}_+ \times S^{2N-3}, \tag{1.26}$$

where

$$S^{2N-3} = \{z \in \mathbb{C}^{N-1}; \sum_{j=1}^{N-1} |z_j|^2 = \sum_{k=1}^{2(N-1)} (q_k)^2 = 1\}.$$

The $U(1)$ action leaves the space $\mathbb{R}_+$ invariant, and is free on S^{2N-3}, so that the factor space (or the shape space) turns out to be isomorphic with

$$\dot{M} = \dot{Q}/U(1) \cong \mathbb{R}_+ \times S^{2N-3}/U(1) \cong \mathbb{R}_+ \times \mathbb{C}P^{N-2}, \tag{1.27}$$

where

$$\mathbb{C}P^{N-2} = S^{2N-3}/U(1)$$

denotes the complex projective space of complex dimension $N-2$, which is defined as the factor space by the equivalence relation defined on S^{2N-3} through $z \sim w \iff z = e^{i\theta}w$. Initially, $\mathbb{C}P^{N-2}$ is defined to be the factor space by the equivalence relation on $\dot{\mathbb{C}}^{N-1}$ through $z \sim w \iff z = \lambda w$.

Proposition 1.1.2 *Except for a singular configuration in which all the particles collide at the center-of-mass, the shape space for the center-of-mass system $\dot{Q}$ is diffeomorphic to $\mathbb{R}_+ \times \mathbb{C}P^{N-2}$.*

What $\mathbb{C}P^{N-2}$ means in the planar many-body system is described as follows: Since $|z|^2 = \sum_{j=1}^{N-1} |z_j|^2 = \sum_{k=1}^{2(N-1)} (q_k)^2 = K(x, x)$, the set S^{2n-3} is viewed as a normalization of molecular configurations, and thereby $\mathbb{C}P^{N-2}$ as the space of normalized shapes of the molecule.

In what follows, we add more explanation of the complex projective space $\mathbb{C}P^{N-2}$, which can be described as the space of $(N-1) \times (N-1)$ complex matrices of the form

$$Z = (z_j \bar{z}_k) = zz^*, \qquad z \in S^{2N-3} \quad j, k = 1, \ldots, N-1. \tag{1.28}$$

We first note that Z is invariant under the $U(1)$ action $(e^{it}z_j\overline{e^{it}z_k} = z_j\overline{z}_k)$. Clearly, the map $[z] \mapsto Z$ is surjective. Conversely, if $Z = (z_j\overline{z}_k) = (w_j\overline{w}_k)$, then a manipulation provides $z = \lambda w$, $|\lambda| = 1$ $(\lambda = (w|z))$, so that $[z] = [w]$, where $(w|z) = \sum_{j=1}^{N-1} \overline{w}_j z_j$ denotes the inner product on $\dot{\mathbb{C}}^{N-1}$. This implies that the map $[z] \mapsto Z$ is injective.

Lemma 1.1.1 *The matrix* $Z = (z_j\overline{z}_k)$ *with* $z \in S^{2N-3} \subset \mathbb{C}^{N-1}$ *is a projection of rank one,*

$$1.\ Z^* = Z, \qquad 2.\ Z^2 = Z, \qquad 3.\ \mathrm{tr}Z = 1. \tag{1.29}$$

Conversely, for an $(N-1)\times(N-1)$ *matrix* Z *satisfying the above properties, there exists a* $z \in S^{2N-3}$ *such that* $Z = (z_j\overline{z}_k)$. *The complex projective space* $\mathbb{C}P^{N-2}$ *is now redefined to be the set of matrices satisfying the above properties.*

Proof It is trivial that Z satisfies the property 1. For the proof of the property 2, one has only to use $\sum |z_j|^2 = 1$. For the proof of the property 3, we note that $\mathrm{tr}Z = \sum |z_j|^2 = 1$.

Conversely, suppose that Z has the properties 1, 2, 3. Then, the property 2 implies that the eigenvalues of Z are 1 and 0. On account of the properties 1 and 3, there exists a unitary matrix U such that

$$U^{-1}ZU = \mathrm{diag}(1, 0, \ldots, 0).$$

Let $(z_1, \ldots, z_{N-1})^T$ denote the first column vector of U. Then, the above equation is written as $Z = (z_j\overline{z}_k)$ with $\sum |z_k|^2 = 1$. This ends the proof.

In concluding this section, we introduce local coordinates of $\dot{Q}$. Let U_k be open subsets of $\dot{\mathbb{C}}^{N-1}$ defined to be

$$U_k = \{z = (z_1, z_2, \cdots, z_{N-1}) \in \dot{\mathbb{C}}^{N-1};\ z_k \neq 0\}, \quad k = 1, 2, \ldots, N-1.$$

In the following, taking $k = N-1$, we deal with U_{N-1}. We introduce local coordinates (θ, r, w^a), $a = 1, 2, \ldots, N-2$, in $U_{N-1} \subset \dot{Q}$ through

$$\begin{aligned} z_a &= re^{i\theta}\rho w_a, \quad a = 1, 2, \ldots, N-2, \\ z_{N-1} &= re^{i\theta}\rho, \end{aligned} \tag{1.30}$$

where

$$r^2 = \sum_{k=1}^{N-1} |z_k|^2, \quad w_a = \frac{z_a}{z_{N-1}}, \quad \rho^{-2} = 1 + \sum_{a=1}^{N-2} |w_a|^2. \tag{1.31}$$

The $U(1)$ action is described in these coordinates as

$$(\theta, r, w_a) \longmapsto (\theta + t, r, w_a), \tag{1.32}$$

which means that (r, w_a) are invariant under the $U(1)$ action, so that they serve as local coordinates of $\pi(U_{N-1})$, which describe the shape of a molecule, and the variable θ describes the attitude of the molecule. In terms of (r, w_a), the matrix elements of Z are expressed, independently of θ, as

$$z_a\overline{z}_b = \frac{r^2 w_a \overline{w}_b}{1 + \sum_{a=1}^{N-2} |w_a|^2}, \qquad z_a\overline{z}_{N-1} = \frac{r^2 w_a}{1 + \sum_{a=1}^{N-2} |w_a|^2}, \qquad etc.$$

1.2 Rotation and Vibration of Planar Many-Body Systems

We give rigorous definitions of rotational and vibrational vectors for planar many-body systems and then proceed to prove that vibrational motions give rise to rotations. Needless to say, we have to distinguish rotations (resp. vibrations) from rotational (resp. vibrational) vectors.

We start with the definition of the tangent space. The tangent space to $\dot{Q}$ at $x \in \dot{Q}$ is defined to be

$$T_x(\dot{Q}) = \{(\boldsymbol{u}_1, \dots, \boldsymbol{u}_N);\ \boldsymbol{u}_\alpha \in \mathbb{R}^2,\ \sum_{\alpha=1}^{N} m_\alpha \boldsymbol{u}_\alpha = 0\}, \quad x = (\boldsymbol{x}_1, \cdots, \boldsymbol{x}_N) \in \dot{Q}, \tag{1.33}$$

where each vector $\boldsymbol{u}_\alpha$ is viewed as a vector at $\boldsymbol{x}_\alpha$ (Fig. 1.4).

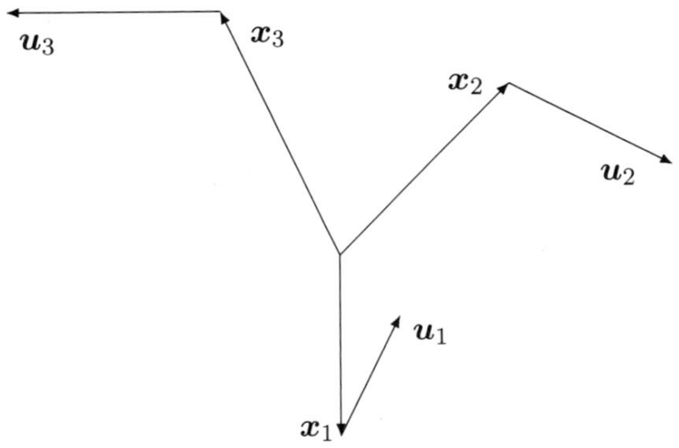

Fig. 1.4 A tangent vector $u = (\boldsymbol{u}_1, \boldsymbol{u}_2, \boldsymbol{u}_3)$ at $x = (\boldsymbol{x}_1, \boldsymbol{x}_2, \boldsymbol{x}_3) \in \dot{Q}$ in the case of a three-body system

The tangent space is endowed with a natural inner product. For tangent vectors $u,\ v \in T_x(\dot{Q})$, the inner product of them is defined to be

$$K_x(u, v) = \sum_{\alpha=1}^{N} m_\alpha \boldsymbol{u}_\alpha \cdot \boldsymbol{v}_\alpha, \quad u = (\boldsymbol{u}_1, \dots, \boldsymbol{u}_N),\ v = (\boldsymbol{v}_1, \dots, \boldsymbol{v}_N), \tag{1.34}$$

where the subscript x of K_x indicates that K_x is defined on the tangent space $T_x(\dot{Q})$. We call K_x the mass-weighted metric (or metric, for short). Let tangent vectors u and v be expressed as $u = \sum U_k f_k$ and $v = \sum V_k f_k$, respectively. Then, the inner product of them is expressed, by the use of (1.13), as

$$K_x(u, v) = \sum_{k=1}^{2(N-1)} U_k V_k. \tag{1.35}$$

We define a rotational vector to be an infinitesimal transformation of the $SO(2)$ action. Since the $SO(2)$ action on the frame f_k is given in (1.20), the derivatives of $g(t) f_k$ with respect to t at $t = 0$ are evaluated as

$$\frac{d}{dt} g(t) f_{2j-1}\bigg|_{t=0} = f_{2j}, \quad \frac{d}{dt} g(t) f_{2j}\bigg|_{t=0} = -f_{2j-1},$$

so that the rotational vector F_x at $x \in \dot{Q}$ is found to be

$$F_x := \frac{d}{dt} g(t) x\bigg|_{t=0} = \sum_{j=1}^{N-1} (-q_{2j} f_{2j-1} + q_{2j-1} f_{2j}), \tag{1.36}$$

where the frame f_k in the above equation is viewed as a frame at $x = \sum_{k=1}^{2(N-1)} q_k f_k$. According to the usual notation in differential geometry, the rotational vector field $x \mapsto F_x$ is expressed as

$$F = \sum_{j=1}^{N-1} \Big(- q_{2j} \frac{\partial}{\partial q_{2j-1}} + q_{2j-1} \frac{\partial}{\partial q_{2j}} \Big). \tag{1.37}$$

To be precise, for any smooth function $f(x)$ on $\dot{Q} \cong \dot{\mathbb{R}}^{2(N-1)}$, F is defined through

$$\begin{aligned} \frac{d}{dt} f(g(t)x)\bigg|_{t=0} &= \sum_{j=1}^{N-1} \Big(\frac{dq_{2j-1}}{dt}(0) \frac{\partial f}{\partial q_{2j-1}} + \frac{dq_{2j}}{dt}(0) \frac{\partial f}{\partial q_{2j}} \Big) \\ &= \sum_{j=1}^{N-1} \Big(- q_{2j} \frac{\partial f}{\partial q_{2j-1}} + q_{2j-1} \frac{\partial f}{\partial q_{2j}} \Big) = (Ff)(x). \end{aligned}$$

Any rotational vector is a scalar multiple of F_x.

A tangent vector $u \in T_x(\dot{Q})$ is called a vibrational vector, if it is orthogonal to all rotational vectors in $T_x(\dot{Q})$. Let $u \in T_x(\dot{Q})$ have the components (U_k) with respect to the frame $\{f_j\}$. Then, on account of (1.36) and (1.35), the u is a vibrational vector, if and only if

$$K_x(u, F_x) = \sum_{j=1}^{N-1}(-U_{2j-1}q_{2j} + U_{2j}q_{2j-1}) = 0. \tag{1.38}$$

Let us introduce complex variables by

$$\zeta_j = U_{2j-1} + iU_{2j}, \qquad i = \sqrt{-1}. \tag{1.39}$$

Then, Eq. (1.38) is rewritten as

$$\sum_{j=1}^{N-1}(-U_{2j-1}q_{2j} + U_{2j}q_{2j-1}) = \frac{1}{2i}\sum_{j=1}^{N-1}(\zeta_j\bar{z}_j - \bar{\zeta}_j z_j) = 0, \tag{1.40}$$

which will be used in the next section. Thus, the tangent space $T_x(\dot{Q})$ is decomposed into the direct sum of the rotational and the vibrational subspaces:

$$T_x(\dot{Q}) = V_{x,\mathrm{rot}} \oplus V_{x,\mathrm{vib}}. \tag{1.41}$$

Now that vibrational vectors are defined, we are in a position to deal with vibrational motions or vibrational curves. A curve $x(t)$ in $\dot{Q}$ is called a vibrational curve, if its tangent vector $\dot{x}(t)$ is always a vibrational vector at $x(t)$. From (1.38), it follows that in the Cartesian coordinates, a curve $x(t) = \sum_{k=1}^{2(N-1)} q_k(t) f_k$ is a vibrational curve, if and only if

$$\sum_{j=1}^{N-1}\left(-q_{2j}\frac{q_{2j-1}}{dt} + q_{2j}\frac{dq_{2j-1}}{dt}\right) = 0, \tag{1.42}$$

which means that the angular momentum of the planar N-body system vanishes.

1.3 Vibrations Induce Rotations in Two Dimensions

Suppose we are given a closed curve $C : \xi(t),\ 0 \le t \le L$, in the shape space $\dot{M}$. For $\xi(0) \in \dot{M}$, there exists a point $x_0 \in \dot{Q}$ such that $\pi(x_0) = \xi(0)$. Then, there exists a vibrational curve $C^* : x(t),\ 0 \le t \le L$, starting at $x_0 = x(0)$ and covering C, $\pi(C^*) = C$. In fact, the fundamental theory of ordinary differential equations ensures the existence of C^* as a solution to (1.42) or to (1.45), as will soon be seen. Though C is a closed curve by definition ($\xi(0) = \xi(L)$), the vibrational curve C^*

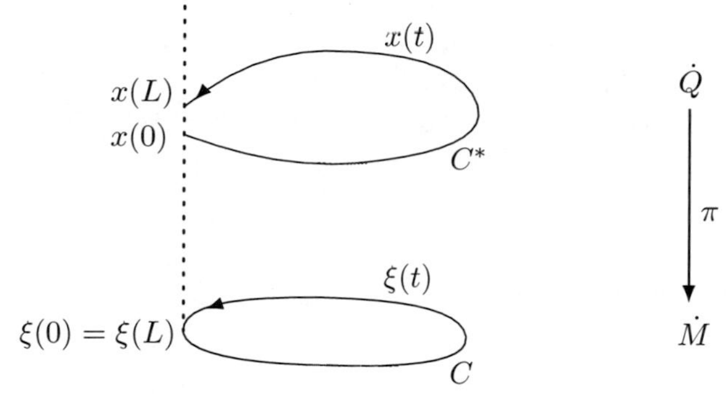

Fig. 1.5 A vibrational motion gives rise to a rotation

is not necessarily closed, $x(0) \neq x(L)$. By the definition of C^*, one has $\pi(x(0)) = \xi(0) = \xi(L) = \pi(x(L))$, which implies that $x(0)$ and $x(L)$ have the same shape as systems of many bodies. However, the conceivable fact that $x(0) \neq x(L)$ means that $x(0)$ and $x(L)$ take different attitudes in the center-of-mass system. Since $x(L)$ and $x(0)$ are equivalent, $[x(L)] = [x(0)]$, there exists a $g \in SO(2)$ such that

$$x(L) = gx(0), \quad g \in SO(2). \tag{1.43}$$

Put another way, the vibrational motion $x(t)$ gives rise to the rotation $g \in SO(2)$ after a cycle of the shape deformation $\xi(t)$, $0 \leq t \leq L$ (Fig. 1.5).

In what follows, we will evaluate rotation angles by using complex local coordinates. Let C be a closed curve in the shape space $\dot{M}$ and $C^* : z(t) \in \dot{Q} \cong \dot{\mathbb{C}}^{N-1}$ its horizontal lift (i.e., a vibrational curve with $\pi(C^*) = C$). Then, from (1.40) with $\dot{z}_j = \zeta_j$, it follows that $z(t)$ should be subject to

$$\sum_{j=1}^{N-1} \left(\frac{dz_j}{dt} \overline{z}_j - \frac{d\overline{z}_j}{dt} z_j \right) = 0. \tag{1.44}$$

In order to evaluate the rotation angle, we adopt the local coordinates introduced in (1.30) with the assumption that $C \subset \pi(U_{N-1})$. Let C^* be expressed as $r = r(t)$, $\theta = \theta(t)$, $w_a = w_a(t)$, $a = 1, \ldots, N-2$, in terms of the local coordinates (r, θ, w_a). Then, on setting $z_a(t) = r(t)e^{i\theta(t)}\rho(t)w_a(t)$, $z_{N-1}(t) = r(t)e^{i\theta(t)}\rho(t)$, Eq. (1.44) is rewritten in terms of (θ, r, w_a) as

$$\frac{d\theta}{dt} + \frac{i}{2} \frac{\displaystyle\sum_{a=1}^{N-2} \left(w_a \frac{d\overline{w}_a}{dt} - \overline{w}_a \frac{dw_a}{dt} \right)}{\displaystyle 1 + \sum_{a=1}^{N-2} |w_a|^2} = 0. \tag{1.45}$$

We note here that the above equation does not include the term dr/dt, which means that $r(t)$ does not contribute a change in the rotation angle. Hence, we may restrict $r(t)$ to a constant value, $r(t) = r_0$, in the closed curve C. Equation (1.45) is easy to integrate. The gain of the rotation angle along the vibrational curve C^* is given by

$$\theta(L) - \theta(0) = \int_0^L \frac{d\theta}{dt} dt = -\frac{i}{2} \int_0^L \frac{\sum_{a=1}^{N-2} \left(w_a \frac{d\overline{w}_a}{dt} - \overline{w}_a \frac{dw_a}{dt} \right)}{1 + \sum_{a=1}^{N-2} |w_a|^2} dt. \tag{1.46}$$

The right-hand side of the above equation is a contour integral along the closed curve C. If we take a variety of closed curves, we can obtain any value of rotation angles. For example, if we take

$$C: \quad w_a(t) = c_a e^{it}, \quad c_a = \text{const.}, \qquad 0 \le t \le 2\pi,$$

then we have the rotation angle

$$\theta(2\pi) - \theta(0) = -\frac{2\pi \sum_{a=1}^{N-2} |c_a|^2}{1 + \sum_{a=1}^{N-2} |c_a|^2}. \tag{1.47}$$

If we reverse the orientation of the curve, the rotation angle given above is reversed in sign. Then, if we choose an orientation of C and the numbers c_a properly, we can obtain any rotation angles from $-\pi$ to π.

Proposition 1.3.1 *The planar many-body system can realize any rotation angle by a suitable vibrational motion. Put another way, for any $g \in SO(2)$, there exists a vibrational curve $x(t)$ satisfying* (1.43).

In the rest of this section, we discuss connection and curvature (see Appendix 5.2 for definition, but the following is comprehensible without referring to it). The tangent space $T_x(\dot{Q})$ is decomposed into the direct sum of the rotational and the vibrational subspaces (see Eq. (1.41)). This decomposition can be described in terms of differential forms. There exists a unique one-form ω satisfying

$$\omega(F) = 1, \quad \omega(v) = 0 \quad \text{for} \quad \forall v \in V_{x,\text{vib}}, \tag{1.48}$$

where F is the rotational vector given in (1.37). The one-form ω is called a connection form. In terms of $(q_1, \cdots, q_{2(N-1)})$ or $(z_1, \cdots, z_{N-1})$, the ω is expressed as

$$
\begin{aligned}
\omega &= \left(\sum_{\ell=1}^{2(N-1)} (q_\ell)^2 \right)^{-1} \sum_{j=1}^{N-1} (q_{2j-1} dq_{2j} - q_{2j} dq_{2j-1}) \\
&= i \left(2 \sum_{j=1}^{N-1} |z_j|^2 \right)^{-1} \sum_{j=1}^{N-1} (z_j d\overline{z}_j - \overline{z}_j dz_j).
\end{aligned}
\tag{1.49}
$$

Further, in terms of (θ, r, w_a) given in (1.30), the ω is written as

$$
\omega = d\theta + \frac{i}{2} \frac{\sum_{a=1}^{N-2} (w_a d\overline{w}_a - \overline{w}_a dw_a)}{1 + \sum_{a=1}^{N-2} |w_a|^2}. \tag{1.50}
$$

For the connection form ω, the curvature form Ω is defined to be $\Omega = d\omega$. A straightforward calculation with (1.49) provides

$$
\Omega = i \left(\sum_{j=1}^{N-1} |z_j|^2 \right)^{-2} \left(\sum_{j=1}^{N-1} |z_j|^2 \sum_{k=1}^{N-1} dz_k \wedge d\overline{z}_k - \sum_{j,k=1}^{N-1} \overline{z}_k z_j dz_k \wedge d\overline{z}_j \right). \tag{1.51}
$$

From (1.50), the Ω proves to take the form

$$
\Omega = i \frac{\left(1 + \sum_{b=1}^{N-2} |w_b|^2 \right) \sum_{a=1}^{N-2} dw_a \wedge d\overline{w}_a - \sum_{a,b=1}^{N-2} \overline{w}_a w_b dw_a \wedge d\overline{w}_b}{\left(1 + \sum_{a=1}^{N-2} |w_a|^2 \right)^2}. \tag{1.52}
$$

We note that the expression of Ω shows that Ω is independent of θ. Then, Ω is viewed as a two-form defined on $\pi(U_{N-1}) \subset \dot{M}$. In conclusion, we remark that by virtue of the Stokes theorem, the rotation angle produced by a vibrational curve covering a closed curve C in $\pi(U_{N-1}) \subset \dot{M}$ is exactly equal to $-\int_S \Omega$, where S is a surface bounded by the closed curve C in $\dot{M}$. The rotation angle obtained in (1.47) is also interpreted in this manner.

1.4 Planar Three-Body Systems

In order to gain a better understanding of rotation, vibration, connection, and curvature, we work with the planar three-body system.

In view of Proposition 1.1.2, we start with a review of $\mathbb{C}P^1$. Lemma 1.1.1 with $N = 3$ provides a realization of $\mathbb{C}P^1$ in the matrix form,

$$\mathbb{C}P^1 \cong \left\{ Z = \begin{pmatrix} |z_1|^2 & z_1\bar{z}_2 \\ z_2\bar{z}_1 & |z_2|^2 \end{pmatrix} \middle| \; |z_1|^2 + |z_2|^2 = 1 \right\}. \tag{1.53}$$

We introduce here the real variables (ξ_1, ξ_2, ξ_3) through

$$\xi_1 + i\xi_2 = 2\bar{z}_1 z_2, \quad \xi_3 = |z_1|^2 - |z_2|^2. \tag{1.54}$$

Then, a straightforward calculation provides

$$\xi_1^2 + \xi_2^2 + \xi_3^2 = (|z_1|^2 + |z_2|^2)^2 = 1, \tag{1.55}$$

which implies that Eq. (1.54) defines a mapping, $Z \mapsto (\xi_1, \xi_2, \xi_3)$, from $\mathbb{C}P^1$ to the unit sphere

$$S^2 = \{(\xi_1, \xi_2, \xi_3) \in \mathbb{R}^3 | \; \xi_1^2 + \xi_2^2 + \xi_3^2 = 1\}.$$

This mapping is injective. In fact, suppose that both $Z = \begin{pmatrix} |z_1|^2 & z_1\bar{z}_2 \\ z_2\bar{z}_1 & |z_2|^2 \end{pmatrix}$ and $W = \begin{pmatrix} |w_1|^2 & w_1\overline{w}_2 \\ w_2\overline{w}_1 & |w_2|^2 \end{pmatrix}$ are mapped to (ξ_1, ξ_2, ξ_3). Then, the equations $\bar{z}_1 z_2 = \overline{w}_1 w_2$ and $|z_1|^2 - |z_2|^2 = |w_1|^2 - |w_2|^2$ imply that there exists $\lambda \in \mathbb{C}$ such that $z_1 = \lambda w_1$, $z_2 = \lambda w_2$ and $|\lambda| = 1$, which results in $Z = W$. Conversely, for any $(\xi_2, \xi_2, \xi_3) \in S^2$, one defines Z to be

$$Z = \frac{1}{2}\begin{pmatrix} 1 + \xi_3 & \xi_1 - i\xi_2 \\ \xi_1 + i\xi_2 & 1 - \xi_3 \end{pmatrix}, \quad \xi_1^2 + \xi_2^2 + \xi_3^2 = 1.$$

Then, it is easy to verify that $Z^* = Z$, $Z^2 = Z$, $\mathrm{tr} Z = 1$ by a simple calculation, which means that $Z \in \mathbb{C}P^1$, and thereby the mapping $Z \mapsto (\xi_1, \xi_2, \xi_3)$ proves to be surjective. Thus, we verify that $\mathbb{C}P^1 \cong S^2$, and therefore, Proposition 1.1.2 with $N = 3$ is rewritten as:

Proposition 1.4.1 *Except for the configuration where three bodies collide at the center-of-mass, the shape space for the planar three-body system is diffeomorphic with* $\mathbb{R}^3 - \{0\}$,

$$\dot{M} \cong \mathbb{R}_+ \times \mathbb{C}P^1 \cong \mathbb{R}_+ \times S^2 \cong \mathbb{R}^3 - \{0\}. \tag{1.56}$$

It then turns out that for the three-body system, the projection π defined in (1.23) is put in the form

$$\pi : \dot{\mathbb{R}}^4 := \mathbb{R}^4 - \{0\} \longrightarrow \dot{\mathbb{R}}^3 := \mathbb{R}^3 - \{0\}. \tag{1.57}$$

We study the map π in more detail. Let (z_1, z_2) denote the complex coordinates of $\mathbb{C}^2$ and q_k, $k = 1, \dots, 4$, the Cartesian coordinates of $\mathbb{R}^4 \cong \mathbb{C}^2$ as before, i.e., $z_1 = q_1 + iq_2$, $z_2 = q_3 + iq_4$. Let (ξ_1, ξ_2, ξ_3) defined in (1.54) be viewed as the Cartesian coordinates of $\mathbb{R}^3$. Under the condition $|z_1|^2 + |z_2|^2 = 1$, the map $(z_1, z_2) \mapsto Z$ defines a map $S^3 \to \mathbb{C}P^1$, and further, the map $Z \mapsto (\xi_1, \xi_2, \xi_3)$ also defines the diffeomorphism $\mathbb{C}P^1 \to S^2$. Accordingly, under the condition $|z_1|^2 + |z_2|^2 = 1$, the composition $(z_1, z_2) \mapsto (\xi_1, \xi_2, \xi_3)$ determines the map $S^3 \to S^2$, which is well known as the Hopf map [12]. Since $\dot{\mathbb{R}}^4 \cong \mathbb{R}_+ \times S^3$ and since $\dot{\mathbb{R}}^3 \cong \mathbb{R}_+ \times S^2$, the map $(z_1, z_2) \mapsto (\xi_1, \xi_2, \xi_3)$ without reference to the condition $|z_1|^2 + |z_2|^2 = 1$ is viewed as giving a map $\dot{\mathbb{R}}^4 \longrightarrow \dot{\mathbb{R}}^3$. Thus, we obtain the explicit expression of π given in (1.57) as

$$\pi : \dot{\mathbb{R}}^4 \longrightarrow \dot{\mathbb{R}}^3; \qquad \begin{cases} \xi_1 = 2\mathrm{Re}(\overline{z}_1 z_2) = 2(q_1 q_3 + q_2 q_4), \\ \xi_2 = 2\mathrm{Im}(\overline{z}_1 z_2) = 2(q_1 q_4 - q_2 q_3), \\ \xi_3 = |z_1|^2 - |z_2|^2 = q_1^2 + q_2^2 - q_3^2 - q_4^2. \end{cases} \tag{1.58}$$

We proceed to rotational and vibrational vectors for the planar three-body system. The tangent space $T_x(\dot{Q})$ has been decomposed into the direct sum of $V_{x,\mathrm{rot}}$ and $V_{x,\mathrm{vib}}$ (see Eq. (1.41)). For the three-body system, those subspaces are explicitly described. By using the orthonormal frame $\{f_j\}$, we define the tangent vectors, F, v_k, $k = 1, 2, 3$, at each point of $\dot{Q} \cong \dot{\mathbb{R}}^4$ to be

$$\begin{aligned} F &= -q_2 f_1 + q_1 f_2 - q_4 f_3 + q_3 f_4, \\ v_1 &= q_3 f_1 + q_4 f_2 + q_1 f_3 + q_2 f_4, \\ v_2 &= q_4 f_1 - q_3 f_2 - q_2 f_3 + q_1 f_4, \\ v_3 &= q_1 f_1 + q_2 f_2 - q_3 f_3 - q_4 f_4, \end{aligned} \tag{1.59}$$

respectively, where F is obtained from (1.36) with $N = 3$, and where we have omitted the subscript $x \in \dot{Q}$. A straightforward calculation shows that these vector fields are orthogonal to one another, which implies that

$$V_{x,\mathrm{rot}} = \mathrm{span}\{F\}, \qquad V_{x,\mathrm{vib}} = \mathrm{span}\{v_1, v_2, v_3\}. \tag{1.60}$$

It is to be noted that the existence of mutually orthogonal vector fields on the whole of $\dot{Q}$ is special to the planar three-body system. In general or in the case of $N > 3$, mutually orthogonal vector fields exist only locally.

So far we have defined rotational and vibrational vectors in a rather abstract way. We now show that the tangent vectors given in (1.59) indeed give rotational and vibrational vectors in a visual manner. To this end, we first recall that the orthonormal system $\{f_j\}$ on $\dot{Q}$ for the three-body system is given, from (1.12), by

$$\begin{aligned} f_1 &= N_1(-m_2\boldsymbol{e}_1, m_1\boldsymbol{e}_1, 0), \\ f_2 &= N_1(-m_2\boldsymbol{e}_2, m_1\boldsymbol{e}_2, 0), \\ f_3 &= N_2(-m_3\boldsymbol{e}_1, -m_3\boldsymbol{e}_1, (m_1+m_2)\boldsymbol{e}_1), \\ f_4 &= N_2(-m_3\boldsymbol{e}_2, -m_3\boldsymbol{e}_2, (m_1+m_2)\boldsymbol{e}_2). \end{aligned}$$

Now we restrict $x = (\boldsymbol{r}_1, \boldsymbol{r}_2) \in \dot{Q}$ to $\boldsymbol{r}_1 = a\boldsymbol{e}_1$, $\boldsymbol{r}_2 = b\boldsymbol{e}_2$ for notational simplicity. Then, by the definition (1.17) of $\boldsymbol{r}_1$, $\boldsymbol{r}_2$, the Cartesian coordinates take the values $q_1 = a$, $q_2 = 0$, $q_3 = 0$, $q_4 = b$, so that Eq. (1.59) reduces to

$$F = af_2 - bf_3, \quad v_1 = bf_2 + af_3, \quad v_2 = bf_1 + af_4, \quad v_3 = af_1 - bf_4.$$

These vectors are illustrated in Fig. 1.6, which can visually explain rotational and vibrational modes of the planar three bodies at the position $(\boldsymbol{r}_1, \boldsymbol{r}_2) = (a\boldsymbol{e}_1, b\boldsymbol{e}_2)$.

We proceed to the connection form. For the planar three-body system, the connection form defined in (1.49) takes the form

$$\omega = r^{-2}(-q_2dq_1 + q_1dq_2 - q_4dq_3 + q_3dq_4), \qquad r^2 = \sum_{k=1}^{4}(q_k)^2. \tag{1.61}$$

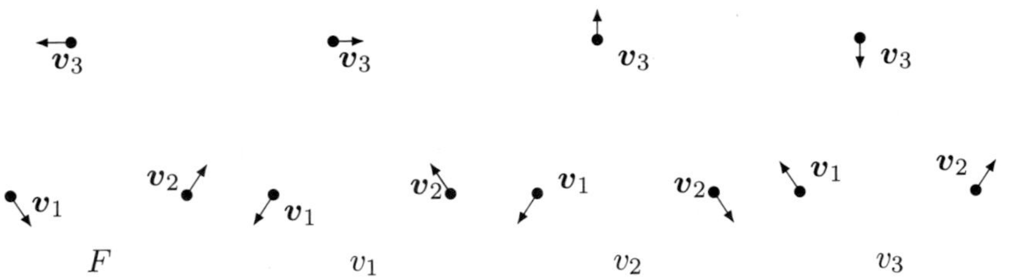

Fig. 1.6 Rotational and vibrational vectors at the configuration $(\boldsymbol{r}_1, \boldsymbol{r}_2) = (a\boldsymbol{e}_1, b\boldsymbol{e}_2)$

Here we view the variables (ξ_k) given in (1.58) as functions on $\dot{\mathbb{R}}^4$ and calculate their differentials to obtain

$$\begin{aligned} d\xi_1 &= 2(q_3dq_1 + q_4dq_2 + q_1dq_3 + q_2dq_4), \\ d\xi_2 &= 2(q_4dq_1 - q_3dq_2 - q_2dq_3 + q_1dq_4), \\ d\xi_3 &= 2(q_1dq_1 + q_2dq_2 - q_3dq_3 - q_4dq_4). \end{aligned} \tag{1.62}$$

We note that the one-forms $\omega, d\xi_k, k = 1, 2, 3$, form a basis of the space of one-forms on $\dot{Q}$. If we consider (ξ_k) as the Cartesian coordinates of $\dot{\mathbb{R}}^3$, we need to use the notation $\pi^* d\xi$, etc., for the left-hand sides of the above equations, where π^* is the symbol for the pull-back of $\pi : \dot{\mathbb{R}}^4 \to \dot{\mathbb{R}}^3$. However, we have omitted the symbol π^* for simplicity.

In a dual manner to $\omega, d\xi_k$, the tangent vectors F, v_k given in (1.59) provide a basis of the space of tangent vector fields, if they are, respectively, expressed as

$$\begin{aligned} F &= -q_2\frac{\partial}{\partial q_1} + q_1\frac{\partial}{\partial q_2} - q_4\frac{\partial}{\partial q_3} + q_3\frac{\partial}{\partial q_4}, \\ v_1 &= q_3\frac{\partial}{\partial q_1} + q_4\frac{\partial}{\partial q_2} + q_1\frac{\partial}{\partial q_3} + q_2\frac{\partial}{\partial q_4}, \\ v_2 &= q_4\frac{\partial}{\partial q_1} - q_3\frac{\partial}{\partial q_2} - q_2\frac{\partial}{\partial q_3} + q_1\frac{\partial}{\partial q_4}, \\ v_3 &= q_1\frac{\partial}{\partial q_1} + q_2\frac{\partial}{\partial q_2} - q_3\frac{\partial}{\partial q_3} - q_4\frac{\partial}{\partial q_4}. \end{aligned} \tag{1.63}$$

It is easy to verify that ω and $d\xi_k/(2r^2)$ are one-forms dual to the vector fields given in (1.63):

$$\omega(F) = 1, \quad \omega(v_k) = 0, \quad d\xi_j(F) = 0, \quad \frac{1}{2r^2}d\xi_j(v_k) = \delta_{jk}, \quad j, k = 1, 2, 3. \tag{1.64}$$

Since these one-forms form a basis of the space of one-forms on $\dot{\mathbb{R}}^4$, the canonical basis dq_k, $k = 1, \ldots, 4$, is expressed in terms of $\omega, d\xi_k$ as

$$\begin{aligned} dq_1 &= -q_2\omega + \frac{1}{2r^2}(q_3d\xi_1 + q_4d\xi_2 + q_1d\xi_3), \\ dq_2 &= q_1\omega + \frac{1}{2r^2}(q_4d\xi_1 - q_3d\xi_2 + q_2d\xi_3), \\ dq_3 &= -q_4\omega + \frac{1}{2r^2}(q_1d\xi_1 - q_2d\xi_2 - q_3d\xi_3), \\ dq_4 &= q_3\omega + \frac{1}{2r^2}(q_2d\xi_1 + q_1d\xi_2 - q_4d\xi_3). \end{aligned} \tag{1.65}$$

We proceed to the curvature form $\Omega = d\omega$. By differentiation of the connection ω given in (1.61), the Ω is written out as

$$\begin{aligned}\Omega = 2r^{-4}[((q_3)^2 + (q_4)^2)dq_1 \wedge dq_2 + ((q_1)^2 + (q_2)^2)dq_3 \wedge dq_4 \\ - (q_2q_3 - q_1q_4)(dq_1 \wedge dq_3 + dq_2 \wedge dq_4) \\ - (q_1q_3 + q_2q_4)(dq_1 \wedge dq_4 - dq_2 \wedge dq_3)].\end{aligned} \tag{1.66}$$

By the use of (1.65), the Ω is rewritten as

$$\Omega = \frac{1}{2}\frac{\xi_1 d\xi_2 \wedge d\xi_3 + \xi_2 d\xi_3 \wedge d\xi_1 + \xi_3 d\xi_1 \wedge d\xi_2}{(\xi_1^2 + \xi_2^2 + \xi_3^2)^{3/2}}. \tag{1.67}$$

Though we have chosen an elementary way to obtain the curvature form shown above, we can also obtain it in a smart manner. In view of (1.63), we can verify that

$$\pi_* v_k = 2r^2 \frac{\partial}{\partial \xi_k}, \quad k = 1, 2, 3. \tag{1.68}$$

In fact, for an arbitrary smooth function f on $\dot{\mathbb{R}}^3$, the defining equation of $\pi_* v_k$ is expressed and arranged as

$$(\pi_* v_k) f = v_k(f \circ \pi) = \sum_{j=1}^{4}\sum_{k=1}^{3} v_k^j \frac{\partial \xi_k}{\partial q_j}\frac{\partial f}{\partial \xi_k} = 2r^2 \frac{\partial f}{\partial \xi_k},$$

where v_k^j are components of $v_k = \sum v_k^j \partial/\partial q_j$. In other words, Eq. (1.68) means that $\frac{1}{2r^2} v_k$ are the horizontal lifts of $\partial/\partial \xi_k$. We denote this fact by

$$\left(\frac{\partial}{\partial \xi_k}\right)^* = \frac{1}{2r^2} v_k, \quad k = 1, 2, 3. \tag{1.69}$$

Further, a straightforward calculation of the brackets among $(\partial/\partial \xi_k)^*$ results in

$$\left[\left(\frac{\partial}{\partial \xi_i}\right)^*, \left(\frac{\partial}{\partial \xi_j}\right)^*\right] = -\sum_k \varepsilon_{ijk} \frac{\xi_k}{2r^6} F, \quad i, j, k = 1, 2, 3. \tag{1.70}$$

Further, we apply the formula $d\omega(X, Y) = X(\omega(Y)) - Y(\omega(X)) - \omega([X, Y])$ (see [75]) for $X = (\partial/\partial \xi_i)^*$ and $Y = (\partial/\partial \xi_j)^*$ to obtain

$$d\omega\left(\left(\frac{\partial}{\partial \xi_i}\right)^*, \left(\frac{\partial}{\partial \xi_j}\right)^*\right) = \frac{1}{2r^6} \sum_k \varepsilon_{ijk} \xi_k,$$

which gives the same result as (1.67).

As was mentioned above, the pull-back symbol π^* is missing in the description of $d\xi_k$ in (1.62). Accordingly, in the right-hand side of (1.67), the pull-back symbol π^* is also missing. However, Eq. (1.67) shows that the Ω can be viewed as a two-form defined on the shape space $\dot{M} \cong \dot{\mathbb{R}}^3$, since Ω is invariant under the action of $SO(2)$. In what follows, we consider Ω as defined on the shape space $\dot{M} \cong \dot{\mathbb{R}}^3$. This curvature has an interesting feature from a physical point of view. Except for the factor $\frac{1}{2}$, the Ω corresponds to the magnetic flux of the Dirac monopole; $\boldsymbol{B} = \boldsymbol{\xi}/|\boldsymbol{\xi}|^3$. In other words, the shape space of the planar three-body system is endowed with a monopole field due to rotation, while this does not mean the existence of the monopole as a particle [26].

We will express the connection and the curvature in other local coordinates. We introduce the coordinates (r, θ, ϕ, ψ) in $\dot{\mathbb{C}}^2$ through

$$z_1 = re^{i\frac{-\phi+\psi}{2}} \cos\frac{\theta}{2}, \quad z_2 = re^{i\frac{\phi+\psi}{2}} \sin\frac{\theta}{2}. \tag{1.71}$$

The range of each variable will be discussed later, but for now it needs to be seen that (θ, ϕ, ψ) serve as coordinates on S^3. The connection form (1.49) with $N = 3$ is expressed, in terms of (1.71), as

$$\omega = \frac{1}{2}(d\psi - \cos\theta d\phi), \tag{1.72}$$

which shows that the present ω is independent of r, so that it can be viewed as a one-form defined on $S^3 \subset \dot{\mathbb{R}}^4$. We proceed to the curvature form to be expressed in the coordinates (θ, ϕ, ψ). Differentiation of ω results in

$$\Omega = d\omega = \frac{1}{2} \sin\theta \, d\theta \wedge d\phi. \tag{1.73}$$

This form can be viewed as $\frac{1}{2}$ times the volume element of S^2 expressed in spherical coordinates. In fact, from (1.54) and (1.71), we obtain

$$\xi_1 = r^2 \sin\theta \cos\phi, \quad \xi_2 = r^2 \sin\theta \sin\phi, \quad \xi_3 = r^2 \cos\theta, \tag{1.74}$$

which implies that (r^2, θ, ϕ) can be viewed as polar spherical coordinates of $\mathbb{R}^3$.

We now return to the question about the ranges of the angular variables given in (1.71). We define the ranges of r, θ, ϕ, ψ to be

$$r > 0, \quad 0 \le \theta \le \pi, \quad 0 \le \phi \le 2\pi, \quad 0 \le \psi \le 4\pi, \tag{1.75}$$

which cover $\dot{\mathbb{C}}^2$ through (1.71). In this coordinate system, the action of $e^{it} \in U(1)$ (see (1.24) with $N = 3$) is expressed as

$$(r, \theta, \phi, \psi) \longmapsto (r, \theta, \phi, \psi + 2t).$$

This implies that the ψ is a coordinate describing the attitude of the planar three bodies, and the r, θ, ϕ are those for the shape, where the ψ ranges over $\mathbb{R}/4\pi\mathbb{Z}$. A further remark is in order. The coordinate system (1.75) fails to work when $\theta = 0, \pi$. In fact, if $\theta = 0$, one has $z_2 = 0$, so that for any $\alpha \in \mathbb{R}$, the pairs (ϕ, ψ) and $(\phi+\alpha, \psi-\alpha)$ correspond to the same z_1. Further, if $\theta = \pi$, then $z_1 = 0$, and hence, for any $\alpha \in \mathbb{R}$, (ϕ, ψ) and $(\phi+\alpha, \psi+\alpha)$ correspond to the same z_2. In spite of this fact, the expression (1.72) of ω is valid, since it has the same value for (ϕ, ψ) and $(\phi+\alpha, \psi \pm \alpha)$.

We will discuss local expressions of ω in more detail. Let U_1 and U_2 denote open subsets of $\dot{\mathbb{C}}^2$ which are defined by $z_1 \neq 0$ and $z_2 \neq 0$, respectively. Then, $(z_1, z_2) \in U_1$ (resp. U_2) if and only if $\theta \neq \pi$ (resp. $\theta \neq 0$). Further, let D_+ and D_- denote the open subsets of S^2 defined to be $D_+ = S^2 - \{S\}$ and $D_- = S^2 - \{N\}$, where S and N stand for the south and the north poles, respectively. The projection $\pi : \dot{\mathbb{R}}^4 \to \dot{\mathbb{R}}^3$ maps U_1 and U_2 to

$$\pi(U_1) \cong \mathbb{R}_+ \times D_+, \qquad \pi(U_2) \cong \mathbb{R}_+ \times D_-,$$

respectively, where $\mathbb{R}_+$ denotes the set of positive real numbers.

Let $\sigma_1 : \pi(U_1) \to U_1$ and $\sigma_2 : \pi(U_2) \to U_2$ denote the local sections defined to be $\sigma_1 : (r^2, \theta, \phi) \mapsto (r^2, \theta, \phi, \phi)$ and $\sigma_2 : (r^2, \theta, \phi) \mapsto (r^2, \theta, \phi, \pi - \phi)$, respectively, or by $\psi = \phi$ and by $\psi = \pi - \phi$, respectively. Then, Eq. (1.72) reduces to

$$\omega_+ = \frac{1}{2}(1 - \cos\theta)d\phi, \qquad \omega_- = \frac{1}{2}(-1 - \cos\theta)d\phi, \tag{1.76}$$

respectively, where $\omega_+ = \sigma_1^*\omega$ and $\omega_- = \sigma_2^*\omega$ denote the value of ω at $\sigma_1(p)$ ($p \in \pi(U_1)$) and $\sigma_2(p)$ ($p \in \pi(U_2)$), respectively. The connection forms $\omega_\pm$ exactly correspond to the locally-defined vector potentials $\boldsymbol{A}_\pm$ for the Dirac monopole $\boldsymbol{B} = (\frac{1}{2})\boldsymbol{\xi}/|\boldsymbol{\xi}|^3$, respectively,

$$\boldsymbol{A}_+ = \frac{-\xi_2\boldsymbol{e}_1 + \xi_1\boldsymbol{e}_2}{2\rho(\rho+\xi_3)}, \qquad \boldsymbol{A}_- = \frac{\xi_2\boldsymbol{e}_1 - \xi_1\boldsymbol{e}_2}{2\rho(\rho-\xi_3)}, \tag{1.77}$$

where $\rho = r^2$ and where $\boldsymbol{A}_\pm$ are defined under the conditions $\rho(\rho \pm \xi_3) \neq 0$, respectively. In fact, we can verify that $\boldsymbol{A}_\pm \cdot d\boldsymbol{\xi} = \omega_\pm$ by straightforward calculation (see also (5.31) for a local description of the connection form).

In the rest of this section, we discuss the metric defined on the shape space for the planar three-body system. Since (q_k) are the Cartesian coordinates with respect to the orthonormal system for the center-of-mass system Q, the canonical (mass-weighted) metric on Q is defined by (1.35) and arranged, by using (1.65), as

$$ds^2 = \sum_{j=1}^{4}(dq_j)^2 = \rho\omega^2 + \frac{1}{4\rho}(d\xi_1^2 + d\xi_2^2 + d\xi_3^2), \quad \rho = |\boldsymbol{\xi}| = r^2, \tag{1.78}$$

which can be also obtained from $K_x(F, F) = r^2$, $K_x(F, v_i) = 0$, $K_x(v_i, v_j) = r^2\delta_{ij}$ and from $\omega(F) = 1$, $d\xi_k(v_j) = 2r^2\delta_{kj}$. The first term of the right-hand side of (1.78) takes non-zero values for vectors in $V_{x,\mathrm{rot}}$, but vanishes for all vectors in $V_{x,\mathrm{vib}}$. The second term vanishes for all the vectors in $V_{x,\mathrm{rot}}$, but takes non-zero values for vectors in $V_{x,\mathrm{vib}}$. Hence, the second term, which is $SO(2)$-invariant, naturally projects onto the shape space $\dot{\mathbb{R}}^3$ to endow $\dot{\mathbb{R}}^3$ with the metric

$$\frac{d\xi_1^2 + d\xi_2^2 + d\xi_3^2}{4\sqrt{\xi_1^2 + \xi_2^2 + \xi_3^2}}. \tag{1.79}$$

Mechanics of the planar three-body system will be discussed in Appendix 5.8.

1.5 The Rotation Group $SO(3)$

In order to study rotation and vibration of many-body systems in three dimensions, we need to give a review of the rotation group $SO(3)$ and its Lie algebra $\mathfrak{so}(3)$. If we put $X \in SO(3)$ in the form $X = (\boldsymbol{x}_1, \boldsymbol{x}_2, \boldsymbol{x}_3)$, where $\boldsymbol{x}_j$ are column vectors in $\mathbb{R}^3$, the conditions $X^T X = I$ with I the 3×3 identity and $\det X = 1$ are rewritten as $\boldsymbol{x}_j \cdot \boldsymbol{x}_k = \delta_{jk}$ and $\det(\boldsymbol{x}_1, \boldsymbol{x}_2, \boldsymbol{x}_2) = 1$, which imply that $\{\boldsymbol{x}_j\}$ form a positively-oriented orthonormal system in $\mathbb{R}^3$.

We are interested in how the set of all positively-oriented orthonormal systems on $\mathbb{R}^3$ is realized. Let $\boldsymbol{x}_j$, $j = 1, 2, 3$, be a positively-oriented orthonormal system. If we move the head of vector $\boldsymbol{x}_1$ freely within the constraint $|\boldsymbol{x}_1| = 1$, it draws the unit sphere S^2. We now parallel translate $\boldsymbol{x}_2$ in $\mathbb{R}^3$ so that its tail may link to the head of $\boldsymbol{x}_1$. Then, the translated $\boldsymbol{x}_2$ becomes a tangent vector to S^2. If we fix $\boldsymbol{x}_1$ arbitrarily and move freely the tangent vector $\boldsymbol{x}_2$ within the restrictions $\boldsymbol{x}_1 \cdot \boldsymbol{x}_2 = 0$, $|\boldsymbol{x}_2| = 1$, the head of $\boldsymbol{x}_2$ draws the unit circle S^1. If $\boldsymbol{x}_1$ and $\boldsymbol{x}_2$ are fixed arbitrarily, the vector $\boldsymbol{x}_3$ is determined by $\boldsymbol{x}_3 = \boldsymbol{x}_1 \times \boldsymbol{x}_2$, which means that $\boldsymbol{x}_3$ is immobile. It then turns out that the set of all the positively-oriented orthonormal systems is realized as the set of all the unit tangent vectors to the unit sphere. This set is called the unit tangent bundle over S^2 and denoted by $T_1(S^2)$ (Fig. 1.7):

$$SO(3) \cong T_1(S^2) = \{(\boldsymbol{x}, \boldsymbol{y}) \in \mathbb{R}^3 \times \mathbb{R}^3;\ |\boldsymbol{x}| = |\boldsymbol{y}| = 1,\ \boldsymbol{x} \cdot \boldsymbol{y} = 0\}. \tag{1.80}$$

We turn to the Lie algebra $\mathfrak{so}(3)$ of $SO(3)$. Let $A \in \mathbb{R}^{3\times 3}$. By definition, A is in $\mathfrak{so}(3)$ if and only if $e^{tA} \in SO(3)$ for $t \in \mathbb{R}$. It then turns out that

$$\mathfrak{so}(3) = \{A \in \mathbb{R}^{3\times 3};\ A + A^T = 0\}. \tag{1.81}$$

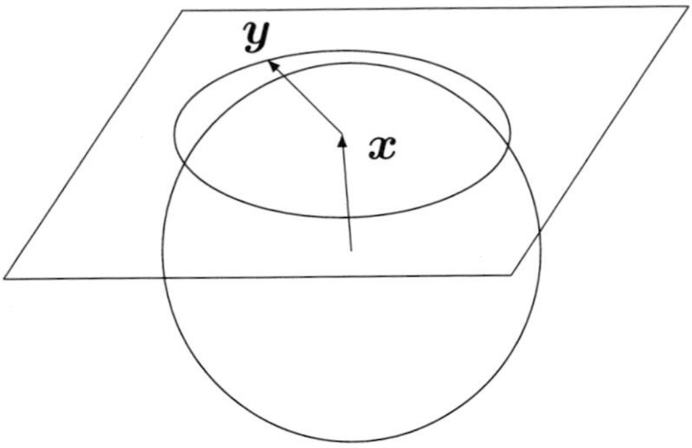

Fig. 1.7 A view of the unit tangent bundle over S^2

The $\mathfrak{so}(3)$ is a real three-dimensional vector space, whose canonical basis is given by

$$L_1 = \begin{pmatrix} 0 & 0 & 0 \\ 0 & 0 & -1 \\ 0 & 1 & 0 \end{pmatrix}, \quad L_2 = \begin{pmatrix} 0 & 0 & 1 \\ 0 & 0 & 0 \\ -1 & 0 & 0 \end{pmatrix}, \quad L_3 = \begin{pmatrix} 0 & -1 & 0 \\ 1 & 0 & 0 \\ 0 & 0 & 0 \end{pmatrix}. \tag{1.82}$$

Then, any $A \in \mathfrak{so}(3)$ is expressed as $A = \sum a_j L_j$, $a_j \in \mathbb{R}$. The isomorphism of $\mathbb{R}^3$ with $\mathfrak{so}(3)$, which is denoted by $R : \mathbb{R}^3 \to \mathfrak{so}(3)$, is given, for $\boldsymbol{a} = \sum a_j \boldsymbol{e}_j \in \mathbb{R}^3$ with $\{\boldsymbol{e}_j\}$ being the standard basis of $\mathbb{R}^3$, by

$$R(\boldsymbol{a}) = \sum_{j=1}^{3} a_j R(\boldsymbol{e}_j) = \sum_{j=1}^{3} a_j L_j = \begin{pmatrix} 0 & -a_3 & a_2 \\ a_3 & 0 & -a_1 \\ -a_2 & a_1 & 0 \end{pmatrix}. \tag{1.83}$$

Proposition 1.5.1 *For $\boldsymbol{a}, \boldsymbol{b} \in \mathbb{R}^3$ and $g \in SO(3)$, the following formulae hold:*

1. $R(\boldsymbol{a})\boldsymbol{x} = \boldsymbol{a} \times \boldsymbol{x}$ for $\boldsymbol{x} \in \mathbb{R}^3$,
2. $\mathrm{Ad}_g R(\boldsymbol{a}) := gR(\boldsymbol{a})g^{-1} = R(g\boldsymbol{a})$,
3. $R(\boldsymbol{a} \times \boldsymbol{b}) = -\boldsymbol{a}\boldsymbol{b}^T + \boldsymbol{b}\boldsymbol{a}^T$.

Proof The proof of the item 1 is clear from the definition of the vector product. To prove the item 2, we use the formula

$$g(\boldsymbol{a} \times \boldsymbol{b}) = g\boldsymbol{a} \times g\boldsymbol{b} \quad \text{for} \quad g \in SO(3), \tag{1.84}$$

which means that the $SO(3)$ action preserves the vector product. This formula can be verified by linearly extending the easy-to-prove formula $g(\boldsymbol{e}_i \times \boldsymbol{e}_j) = g\boldsymbol{e}_k = g\boldsymbol{e}_i \times g\boldsymbol{e}_j$, where (i, j, k) is a cyclic permutation of $(1, 2, 3)$. Then, we find that $gR(\boldsymbol{a})g^{-1}\boldsymbol{x} = g(\boldsymbol{a} \times g^{-1}\boldsymbol{x}) = g\boldsymbol{a} \times \boldsymbol{x} = R(g\boldsymbol{a})\boldsymbol{x}$ for all $\boldsymbol{x}$. The item 3 is a consequence of the formula

$$\boldsymbol{a} \times (\boldsymbol{b} \times \boldsymbol{c}) = (\boldsymbol{a} \cdot \boldsymbol{c})\boldsymbol{b} - (\boldsymbol{a} \cdot \boldsymbol{b})\boldsymbol{c}. \tag{1.85}$$

In fact, by using this formula, we find that $R(\boldsymbol{a} \times \boldsymbol{b})\boldsymbol{x} = (-\boldsymbol{a}\boldsymbol{b}^T + \boldsymbol{b}\boldsymbol{a}^T)\boldsymbol{x}$. This ends the proof.

The $\mathfrak{so}(3)$ is endowed with the inner product through

$$\langle A, B\rangle = \frac{1}{2}\mathrm{tr}(A^T B), \quad A,\ B \in \mathfrak{so}(3). \tag{1.86}$$

Proposition 1.5.2 *With respect to the inner product* (1.86)*, the* L_j, $j = 1, 2, 3$, *form an orthonormal system of* $\mathfrak{so}(3)$*, and further* $\mathfrak{so}(3)$ *and* $\mathbb{R}^3$ *are isometric:*

1. $\langle L_j, L_k\rangle = \delta_{jk}, \qquad j, k = 1, 2, 3,$
2. $\langle R(\boldsymbol{a}), R(\boldsymbol{b})\rangle = \boldsymbol{a} \cdot \boldsymbol{b} \qquad$ for $\quad \boldsymbol{a},\ \boldsymbol{b} \in \mathbb{R}^3.$

The proof is easily performed by using the definition of the inner product.

Proposition 1.5.3 *As Lie algebras,* $\mathfrak{so}(3) \cong \mathbb{R}^3$ *have the following properties:*

1. $[L_j, L_k] = \sum_{\ell=1}^{3} \varepsilon_{jk\ell} L_\ell, \quad \boldsymbol{e}_j \times \boldsymbol{e}_k = \sum_{\ell=1}^{3} \varepsilon_{jk\ell}\boldsymbol{e}_\ell,$
2. $[R(\boldsymbol{a}), R(\boldsymbol{b})] = R(\boldsymbol{a} \times \boldsymbol{b}), \quad \boldsymbol{a},\ \boldsymbol{b} \in \mathbb{R}^3,$
3. $\langle R(\boldsymbol{a}), [R(\boldsymbol{b}), R(\boldsymbol{c})]\rangle = \boldsymbol{a} \cdot (\boldsymbol{b} \times \boldsymbol{c}), \quad \boldsymbol{a}, \boldsymbol{b}, \boldsymbol{c} \in \mathbb{R}^3.$

Proof The item 1 is easy to prove by calculation. To prove the item 2, we use the formula (1.85). For any $\boldsymbol{x} \in \mathbb{R}^3$, we can show that $[R(\boldsymbol{a}), R(\boldsymbol{b})]\boldsymbol{x}$ and $R(\boldsymbol{a} \times \boldsymbol{b})\boldsymbol{x}$ coincide by using the item 1 of Proposition 1.5.1 together with the formula (1.85). Note that since $L_i = R(\boldsymbol{e}_i)$, the first equation of the item 1 is a consequence the item 2 and the second equation of the item 1. The item 3 is a consequence of the item 2 of the present proposition and the item 2 of Proposition 1.5.2. This ends the proof.

Applying the definition, $\exp(tA) = \sum_{n=0}^{\infty} \frac{t^n}{n!} A^n$, of matrix exponential, we obtain

$$\exp(tR(\boldsymbol{e}_1)) = \begin{pmatrix} 1 & 0 & 0 \\ 0 & \cos t & -\sin t \\ 0 & \sin t & \cos t \end{pmatrix},$$

$$\exp(tR(\boldsymbol{e}_2)) = \begin{pmatrix} \cos t & 0 & \sin t \\ 0 & 1 & 0 \\ -\sin t & 0 & \cos t \end{pmatrix}, \tag{1.87}$$

$$\exp(tR(\boldsymbol{e}_3)) = \begin{pmatrix} \cos t & -\sin t & 0 \\ \sin t & \cos t & 0 \\ 0 & 0 & 1 \end{pmatrix}.$$

We can obtain the same results by solving associated differential equations. For example, to find the matrix expression of $\exp(tR(\boldsymbol{e}_2))$, we have only to solve the linear differential equation $d\boldsymbol{x}/dt = R(\boldsymbol{e}_2)\boldsymbol{x}, \ \boldsymbol{x} \in \mathbb{R}^3$. In fact, the coupled first-order differential equations reduce to the uncoupled second-order differential equations $d^2x_3/dt^2 = -x_3$ etc, which are easy to solve. We then find that $x_1(t) = x_3(0)\sin t + x_1(0)\cos t, \ x_2(t) = x_2(0), \ x_3(t) = x_3(0)\cos t - x_1(0)\sin t$. Putting these solutions in the form $\boldsymbol{x}(t) = M(t)\boldsymbol{x}(0)$ and comparing it with the generic expression of the solution, $\boldsymbol{x}(t) = \exp(tR(\boldsymbol{e}_2))\boldsymbol{x}(0)$, we find that $M(t) = \exp(tR(\boldsymbol{e}_2))$ on account of the uniqueness of solution.

Clearly, $\exp(tR(\boldsymbol{e}_j))$ describes a rotation about the $\boldsymbol{e}_j$-axis with $j = 1, 2, 3$. (Though the axis of the rotation is a line which is fixed under the rotation, we use the same word for a vector along the axis.) In fact, one has $\exp(tR(\boldsymbol{e}_j))\boldsymbol{e}_j = \boldsymbol{e}_j$, which means that the action of $\exp(tR(\boldsymbol{e}_j))$ fixes the vector $\boldsymbol{e}_j$. In general, the following lemma is valid for rotation.

Proposition 1.5.4 *Let* $\boldsymbol{a}$ *be a unit vector* ($|\boldsymbol{a}| = 1$). *Then,* $\exp(tR(\boldsymbol{a}))$ *is a rotation about the axis* $\boldsymbol{a}$. *Conversely, any rotation about the axis* $\boldsymbol{a}$ *is expressed as* $\exp(tR(\boldsymbol{a}))$.

Proof It is easy to see that $\exp(tR(\boldsymbol{a})) \in O(3)$. Since $\det\exp(tR(\boldsymbol{a})) = \exp(t\mathrm{tr}(R(\boldsymbol{a}))) = 1$, where the symbol exp in the right-hand side stands for the scalar exponential, we find that $\exp(tR(\boldsymbol{a})) \in SO(3)$. From $R(\boldsymbol{a})\boldsymbol{a} = \boldsymbol{a} \times \boldsymbol{a} = 0$, it follows that $\exp(tR(\boldsymbol{a}))\boldsymbol{a} = \boldsymbol{a}$. Conversely, to express a rotation about the axis $\boldsymbol{a}$, we introduce mutual orthogonal unit vectors $\boldsymbol{b}$ and $\boldsymbol{c}$ which are also orthogonal to $\boldsymbol{a}$ in such a manner that the system $\boldsymbol{b}, \boldsymbol{c}, \ \boldsymbol{a}$ form a positively-oriented orthonormal system. Then, a rotation g about the axis $\boldsymbol{a}$ is put, with respect to the basis $\{\boldsymbol{b}, \boldsymbol{c}, \boldsymbol{a}\}$, in the form

$$g\boldsymbol{b} = \boldsymbol{b}\cos t + \boldsymbol{c}\sin t,$$

$$g\boldsymbol{c} = -\boldsymbol{b}\sin t + \boldsymbol{c}\cos t,$$

$$g\boldsymbol{a} = \boldsymbol{a}.$$

These are also expressed as

$$(g\boldsymbol{b}, g\boldsymbol{c}, g\boldsymbol{a}) = (\boldsymbol{b}, \boldsymbol{c}, \boldsymbol{a}) \begin{pmatrix} \cos t & -\sin t & 0 \\ \sin t & \cos t & 0 \\ 0 & 0 & 1 \end{pmatrix}.$$

Here we denote by h the matrix whose column vectors are $\boldsymbol{b}, \ \boldsymbol{c}, \ \boldsymbol{a}$. It is clear that $h \in SO(3)$ and $h\boldsymbol{e}_3 = \boldsymbol{a}$. Then, the above equation takes the form $gh = h\exp(tR(\boldsymbol{e}_3))$, so that one has

$$g = he^{tR(\boldsymbol{e}_3)}h^{-1} = e^{thR(\boldsymbol{e}_3)h^{-1}} = e^{tR(h\boldsymbol{e}_3)} = e^{tR(\boldsymbol{a})},$$

where use has been made of the formula 2 of Proposition 1.5.1. This ends the proof.

We now describe the rotation $e^{tR(\boldsymbol{a})}$ in an explicit form. We first note that $e^{tR(\boldsymbol{a})}\boldsymbol{a} = \boldsymbol{a}$, which means that $\boldsymbol{a}$ is on the axis of rotation. For any unit vector $\boldsymbol{a}$, there exists a matrix $g \in SO(3)$ such that $g\boldsymbol{e}_3 = \boldsymbol{a}$. Then, $e^{tR(\boldsymbol{a})}$ is arranged as

$$e^{tR(\boldsymbol{a})} = e^{tR(g\boldsymbol{e}_3)} = g e^{tR(\boldsymbol{e}_3)} g^{-1}.$$

We bring $e^{tR(\boldsymbol{e}_3)}$ into the form

$$e^{tR(\boldsymbol{e}_3)} = \cos t \left(I - \begin{pmatrix} 0 & 0 & 0 \\ 0 & 0 & 0 \\ 0 & 0 & 1 \end{pmatrix} \right) + \sin t\, R(\boldsymbol{e}_3) + \begin{pmatrix} 0 & 0 & 0 \\ 0 & 0 & 0 \\ 0 & 0 & 1 \end{pmatrix}.$$

Operating the both sides of the above equation with Ad_g and using the formula

$$\boldsymbol{e}_3 \boldsymbol{e}_3^T = \begin{pmatrix} 0 & 0 & 0 \\ 0 & 0 & 0 \\ 0 & 0 & 1 \end{pmatrix},$$

we obtain

$$\begin{aligned} e^{tR(\boldsymbol{a})} &= \cos t \big(I - g\boldsymbol{e}_3 (g\boldsymbol{e}_3)^T \big) + \sin t R(g\boldsymbol{e}_3) + g\boldsymbol{e}_3 (g\boldsymbol{e}_3)^T \\ &= \cos t (I - \boldsymbol{a}\boldsymbol{a}^T) + \sin t\, R(\boldsymbol{a}) + \boldsymbol{a}\boldsymbol{a}^T, \end{aligned} \tag{1.88}$$

which is an explicit expression of the rotation about a unit vector $\boldsymbol{a}$.

We can prove further the following proposition on the rotation axis.

Proposition 1.5.5 *Any rotation $g \in SO(3)$ has a rotation axis. In particular, if $g \neq I$, then the rotation axis for g is unique.*

Proof Since

$$\det(g - I) = \det(g^T - I) = \det(g^T) \det(I - g) = -\det(g - I),$$

we see that $\det(g - I) = 0$. Then, there exists an eigenvector $\boldsymbol{x} \neq 0$ associated with the eigenvalue 1; $g\boldsymbol{x} = \boldsymbol{x}$, which means that the line containing $\boldsymbol{x}$ is a rotation axis. Suppose that there are two axes for g, i.e., there exist mutually independent $\boldsymbol{x}$ and $\boldsymbol{y}$ such that $g\boldsymbol{x} = \boldsymbol{x},\ g\boldsymbol{y} = \boldsymbol{y}$. Then, it follows that $g(\boldsymbol{x} \times \boldsymbol{y}) = g\boldsymbol{x} \times g\boldsymbol{y} = \boldsymbol{x} \times \boldsymbol{y}$. This implies that g leaves invariant three mutually independent vectors $\boldsymbol{x},\ \boldsymbol{y},\ \boldsymbol{x} \times \boldsymbol{y}$ of $\mathbb{R}^3$, so that it leaves all the vectors of $\mathbb{R}^3$ invariant, which means that $g = I$. Therefore, the rotation axis should be unique, if $g \neq I$. This ends the proof.

So far we have shown that any rotation has a rotation axis and that a rotation about the axis $\boldsymbol{a}$ is put in the form $\exp(tR(\boldsymbol{a}))$ with $|\boldsymbol{a}| = 1$. Because of the periodicity of rotation, the totality of rotations forms the set

$$B^3 = \{\boldsymbol{x} \in \mathbb{R}^3;\ |\boldsymbol{x}| \leq \pi\}.$$

In addition, on account of (1.88), every pair of antipodal points on the boundary of B^3 should be identified. The B^3 with the present identification, which we denote by $\hat{B}^3$, can be shown to be isomorphic with the real projective space $\mathbb{R}P^3$, where $\mathbb{R}P^3$ is defined to be the factor space of S^3 by identification of pairs of antipodal points. Let $S^3(\pi)$ denote the three-sphere of radius π, which is realized in $\mathbb{R}^4$ by the condition $|\boldsymbol{y}| = \pi$, $\boldsymbol{y} \in \mathbb{R}^4$. The real projective space $\mathbb{R}P^3$ is realized through the identification of the antipodal points, $\boldsymbol{y}$ and $-\boldsymbol{y}$, of $S^3(\pi)$. Put another way, $\mathbb{R}P^3$ is the factor space, $S^3(\pi)/\mathbb{Z}_2$, by the action of the group $\mathbb{Z}_2 = \{\pm 1\}$. We denote by $[\boldsymbol{y}]$ the equivalence class of $\boldsymbol{y} \in S^3(\pi)$, so that $[-\boldsymbol{y}] = [\boldsymbol{y}]$. For $[\boldsymbol{y}] \in \mathbb{R}P^3$ with $y_4 \neq 0$, either $\boldsymbol{y}$ or $-\boldsymbol{y}$ has a negative fourth coordinate. We assume that $\boldsymbol{y}$ has $y_4 < 0$, without loss of generality. We draw a line segment joining the north pole $(0, 0, 0, \pi) \in S^3(\pi)$ and the $\boldsymbol{y} \in S^3(\pi)$ in question. Let $(x_1, x_2, x_3, 0)$ be the point at which the line segment crosses the plane $y_4 = 0$ in $\mathbb{R}^4$. Then, we have

$$x_k = \frac{\pi y_k}{\pi - y_4}, \qquad k = 1, 2, 3.$$

Let us denote these points by $\boldsymbol{x} \in \mathbb{R}^3 \subset \mathbb{R}^4$ with $|\boldsymbol{x}| < \pi$, which correspond to the interior points of B^3. If $y_4 = 0$, then $|\boldsymbol{y}| = \pi$ defines the two-sphere $S^2(\pi)$ of radius π, corresponding to the boundary of B^3. Since $[\boldsymbol{y}] \in \mathbb{R}P^3$ and since $S^2(\pi)$ in question is on the equator of $S^3(\pi)$, antipodal points of $S^2(\pi)$ should be identified. Thus, we have constructed the one-to-one correspondence between $\mathbb{R}P^3$ and $\hat{B}^3$.

We can show that any $g \in SO(3)$ is composed of rotations about only two axes.

Proposition 1.5.6 *Let $\boldsymbol{e}_j$, $j = 1, 2, 3$, be the standard basis of $\mathbb{R}^3$. (We may take any orthonormal system for $\boldsymbol{e}_j$.) Any $g \in SO(3)$ can be generated by rotations about chosen two axes, say $\boldsymbol{e}_2$, $\boldsymbol{e}_3$.*

Proof The vector $\boldsymbol{e}_3$ is transformed to $g\boldsymbol{e}_3$ by the action of $g \in SO(3)$. We carry $g\boldsymbol{e}_3$ to a vector in the $\boldsymbol{e}_3$–$\boldsymbol{e}_1$ plane by a rotation h_3 about the $\boldsymbol{e}_3$ axis and further move the vector in question to $\boldsymbol{e}_3$ by a rotation about the $\boldsymbol{e}_2$ axis. This procedure provides $h_2 h_3 g\boldsymbol{e}_3 = \boldsymbol{e}_3$. This implies that $h_2 h_3 g$ is a rotation about the $\boldsymbol{e}_3$ axis, which we denote by k_3, so that we have $h_2 h_3 g = k_3$. Hence, we obtain $g = h_3^{-1} h_2^{-1} k_3$, as is wanted.

In the course of the proof, we can restrict the rotation angle of h_3 to $-2\pi \leq -\phi \leq 0$ so that $h_3 g\boldsymbol{e}_3$ may be sitting on the half of the $\boldsymbol{e}_3$–$\boldsymbol{e}_1$ plane with $+\boldsymbol{e}_1$. Then, the rotation angle of h_2 about the $\boldsymbol{e}_2$ axis may take the range $-\pi \leq -\theta \leq 0$. The rotation angle of k_3 about the $\boldsymbol{e}_3$ axis has the range $0 \leq \psi \leq 2\pi$ of course. Thus we have obtained the following theorem.

Theorem 1.5.1 *For any $g \in SO(3)$, there are three angles $\phi,\ \psi,\ \theta$ such that*

$$g = e^{\phi R(e_3)} e^{\theta R(e_2)} e^{\psi R(e_3)},$$
$$0 \le \phi \le 2\pi, \quad 0 \le \theta \le \pi, \quad 0 \le \psi \le 2\pi. \tag{1.89}$$

These angles are called the Euler angles.

We should be careful about the ranges of the Euler angles. If $\theta \neq 0,\ \pi$, the Euler angles are uniquely determined, but if $\theta = 0,\ \pi$, they are not. To see this, we denote by $g(\phi, \theta, \psi)$ the rotation with the Euler angles $(\phi,\ \theta,\ \psi)$. If $g(\phi, \theta, \psi)$ has other Euler angles (ϕ', θ', ψ'), the equation $g(\phi, \theta, \psi)\boldsymbol{e}_3 = g(\phi', \theta', \psi')\boldsymbol{e}_3$ provides $\phi = \phi',\ \theta = \theta'$ if $\theta \neq 0,\ \pi$. Further, the equation $g(\phi, \theta, \psi) = g(\phi, \theta, \psi')$ results in $\psi = \psi'$. In contrast to this, if $\theta = 0,\ \pi$, the easy-to-prove equations $e^{\alpha R(e_3)} e^{-\alpha R(e_3)} = I$ and $e^{\alpha R(e_3)} e^{\pi R(e_2)} e^{\alpha R(e_3)} = e^{\pi R(e_2)}$ with $\alpha \in \mathbb{R}$ imply that $g(\phi+\alpha, 0, \psi-\alpha) = g(\phi, 0, \psi)$ and $g(\phi+\alpha, \pi, \psi+\alpha) = g(\phi, \pi, \psi)$, respectively.

Further remarks are in order. As is stated in Proposition 1.5.6, the choice of the rotation axes is not unique, so that the Euler angles can be defined in various ways. One may choose $\boldsymbol{e}_1,\ \boldsymbol{e}_3$ as rotation axes. In this case, the procedure to obtain the Euler angles is as follows: We carry the vector $g\boldsymbol{e}_3$ to a vector in the $\boldsymbol{e}_2$–$\boldsymbol{e}_3$ plane by a rotation about the $\boldsymbol{e}_3$ axis to get $h_3 g\boldsymbol{e}_3$, and further translate it to $\boldsymbol{e}_3$ by a rotation about the $\boldsymbol{e}_1$ axis to obtain $h_1 h_3 g\boldsymbol{e}_3 = \boldsymbol{e}_3$. Then, we have $g = h_3^{-1} h_1^{-1} k_3$. In this case, Eq. (1.89) takes the form $g = e^{\phi\widehat{e}_3} e^{\theta\widehat{e}_1} e^{\psi\widehat{e}_3}$, where $\widehat{\boldsymbol{e}}_i = R(\boldsymbol{e}_i)$. If the two axes $\boldsymbol{e}_1, \boldsymbol{e}_2$ are chosen, the same reasoning results in $g = e^{\phi\widehat{e}_2} e^{\theta\widehat{e}_1} e^{\psi\widehat{e}_2}$. Furthermore, we may take three vectors $\boldsymbol{e}_j$ as rotation axes. A procedure for the Euler angles is as follows: For any $g \in SO(3)$, we first take $g\boldsymbol{e}_1$. By $h_2 \in SO(3)$, we rotate $g\boldsymbol{e}_1$ about $\boldsymbol{e}_2$ to put it on the plane $\boldsymbol{e}_1$–$\boldsymbol{e}_2$ with $+\boldsymbol{e}_1$. We then rotate the vector $h_2 g\boldsymbol{e}_1$ by a rotation h_3 about $\boldsymbol{e}_3$ to place it on $\boldsymbol{e}_1$. Then, we have $h_3 h_2 g\boldsymbol{e}_1 = \boldsymbol{e}_1$. This means that $h_3 h_2 g$ is a rotation about $\boldsymbol{e}_1$, which we denote by k_1. Thus, the g proves to expressed as $g = h_2^{-1} h_3^{-1} k_1$. If we denote by χ_a, $a = 1, 2, 3$, three angles associated with h_2^{-1}, h_3^{-1}, k_1, respectively, we obtain

$$g = e^{\chi_2 R(e_2)} e^{\chi_3 R(e_3)} e^{\chi_1 R(e_1)}, \tag{1.90}$$
$$0 \le \chi_1 \le 2\pi, \quad 0 \le \chi_2 \le 2\pi, \quad -\frac{\pi}{2} \le \chi_3 \le \frac{\pi}{2},$$

which will be used in the falling cat problem.

We now explain a geometric interpretation of the Euler angles $(\phi,\ \theta,\ \psi)$ given in (1.89) from the viewpoint of the identification $SO(3) = T_1(S^2)$ given in (1.80). The angles $(\phi,\ \theta)$ of the Euler angles give rise to the spherical coordinates of S^2. In fact, from $|g\boldsymbol{e}_3| = 1$, we obtain

$$S^2 = \{g\boldsymbol{e}_3;\ g \in SO(3)\},$$

and the vector ge_3 is expressed as

$$ge_3 = e_1 \sin\theta \cos\phi + e_2 \sin\theta \sin\phi + e_3 \cos\theta, \quad 0 \le \theta \le \pi,\ 0 \le \phi \le 2\pi.$$

To see a geometric interpretation of the angular variable ψ, we put the vector ge_1 in the form

$$ge_1 = u_1 \cos\psi + u_2 \sin\psi, \tag{1.91}$$

where

$$u_1 = e^{\phi R(e_3)} e^{\theta R(e_2)} e_1 = e_1 \cos\theta \cos\phi + e_2 \cos\theta \sin\phi - e_3 \sin\theta,$$

$$u_2 = e^{\phi R(e_3)} e^{\theta R(e_2)} e_2 = -e_1 \sin\phi + e_2 \cos\phi.$$

As is easily seen, the vectors u_1 and u_2 are mutually orthogonal unit vectors, which are perpendicular to ge_3. We parallel translate u_1 and u_2 in $\mathbb{R}^3$ so that their tails link to the head of ge_3. Then, the translated vectors are viewed as an orthonormal system on the tangent plane to S^2. From this point of view, Eq. (1.91) implies that the variable ψ is an angle variable of the unit circle drawn by ge_1 on the tangent plane to S^2 at ge_3 with $(\phi,\ \theta)$ fixed. Put another way, ψ $(0 \le \psi \le 2\pi)$ is an angle variable for the unit circle over the sphere S^2. We remark here that the vectors $x,\ y$ in the definition (1.80) of $T_1(S^2)$ can be taken as ge_3 and ge_1, respectively. We do not have to talk about ge_2, since we have $ge_3 \times ge_1 = ge_2$.

In what follows, we describe the left-invariant one-forms and right-invariant one-forms and their dual vector fields, which are typically used in mechanics for rigid bodies. For $g \in SO(3)$ viewed as a matrix variable, the one-forms

$$g^{-1}dg, \quad dgg^{-1} \tag{1.92}$$

are called the left- and the right-invariant (one-)forms, respectively. In fact, for a constant matrix $h \in SO(3)$, we verify that

$$(hg)^{-1}d(hg) = g^{-1}dg, \quad d(gh)(gh)^{-1} = dgg^{-1},$$

which means that the forms $g^{-1}dg$ and dgg^{-1} are invariant under the left and the right translations, $g \mapsto hg$ and $g \mapsto gh$, respectively. Differentiating $g^{-1}g = I$, we obtain $dg^{-1}g + g^{-1}dg = (g^{-1}dg)^T + g^{-1}dg = 0$, which implies that $g^{-1}dg$ is anti-symmetric. In a similar manner, we see that dgg^{-1} is anti-symmetric as well. We denote by Φ^a and Ψ^a the components of the right- and the left-invariant one-forms, respectively:

$$dgg^{-1} = \sum_{a=1}^{3} \Phi^a R(e_a), \quad g^{-1}dg = \sum_{a=1}^{3} \Psi^a R(e_a). \tag{1.93}$$

A straightforward calculation in terms of the Euler angles can provide

$$\begin{cases} \Phi^1 = -\sin\phi d\theta + \sin\theta\cos\phi d\psi, \\ \Phi^2 = \cos\phi d\theta + \sin\theta\sin\phi d\psi, \\ \Phi^3 = d\phi + \cos\theta d\psi, \end{cases} \tag{1.94}$$

$$\begin{cases} \Psi^1 = \sin\psi d\theta - \sin\theta\cos\psi d\phi, \\ \Psi^2 = \cos\psi d\theta + \sin\theta\sin\psi d\phi, \\ \Psi^3 = d\psi + \cos\theta d\phi. \end{cases} \tag{1.95}$$

To show this, we first put the rotation matrix g in the form

$$g = g_3(\phi)g_2(\theta)g_3(\psi),$$

as is shown in (1.89), where g_3, g_2 denote rotations about the $\boldsymbol{e}_3$- and the $\boldsymbol{e}_2$-axes, respectively. For the factor matrices, one verifies that

$$dg_3(\phi)g_3^{-1}(\phi) = R(\boldsymbol{e}_3)d\phi, \quad dg_2(\theta)g_2^{-1}(\theta) = R(\boldsymbol{e}_2)d\theta.$$

It then follows that dgg^{-1} is expressed as

$$dgg^{-1} = R(\boldsymbol{e}_3)d\phi + R(g_3(\phi)\boldsymbol{e}_2)d\theta + R(g_3(\phi)g_2(\theta)\boldsymbol{e}_3)d\psi.$$

Further, by using

$$g_2(\theta)\boldsymbol{e}_3 = \boldsymbol{e}_3\cos\theta + \boldsymbol{e}_1\sin\theta,$$

$$g_3(\phi)\boldsymbol{e}_1 = \boldsymbol{e}_1\cos\phi + \boldsymbol{e}_2\sin\phi,$$

$$g_3(\phi)\boldsymbol{e}_2 = -\boldsymbol{e}_1\sin\phi + \boldsymbol{e}_2\cos\phi,$$

we can obtain the explicit expression (1.94) of Φ^a, $a = 1, 2, 3$. A similar method can be applied to find the expression (1.95) of Ψ^a.

From the definition (1.92) of the left- and right-invariant one-forms, we find, by differentiation, that

$$d(g^{-1}dg) = -g^{-1}dg \wedge g^{-1}dg, \quad d(dgg^{-1}) = dgg^{-1} \wedge dgg^{-1},$$

which are written out to give, respectively,

$$d\Psi^c = -\frac{1}{2}\sum \varepsilon_{abc}\Psi^a \wedge \Psi^b, \quad d\Phi^c = \frac{1}{2}\sum \varepsilon_{abc}\Phi^a \wedge \Phi^b. \tag{1.96}$$

In a dual manner to the left- and the right-invariant one-forms, we can define left- and right-invariant vector fields on $SO(3)$. Let $h(t)$ be a one-parameter subgroup of $SO(3)$. Then, the vector fields defined to be

$$X_g = \frac{d}{dt} g h(t)\Big|_{t=0}, \quad Y_g = \frac{d}{dt} h(t) g\Big|_{t=0}, \tag{1.97}$$

are called a left- and a right-invariant vector field, respectively. If $h(t) = e^{tR(\boldsymbol{a})}$, one obtains

$$X_g = gR(\boldsymbol{a}), \quad Y_g = R(\boldsymbol{a})g,$$

respectively. In what follows, we will find the left- and the right-invariant vector fields in terms of the Euler angles, which are dual to Ψ^a and to Φ^a, respectively.

For the vector $\boldsymbol{\Psi} = \sum \Psi^a \boldsymbol{e}_a$ of the left-invariant one-forms, we have $dg = gR(\boldsymbol{\Psi})$. In particular, for a curve $g(t)$ passing g at $t = 0$, this equation reduces to $dg/dt = gR(\dot{\boldsymbol{\Psi}})$, where $\dot{\boldsymbol{\Psi}} = \boldsymbol{\Psi}(d/dt)$, i.e., $\dot{\Psi}_1 = \sin\psi\,\dot{\theta} - \sin\theta\cos\psi\,\dot{\phi}$ etc. For $g(t) = ge^{tR(\boldsymbol{a})}$, the present equation is evaluated to give $gR(\boldsymbol{a}) = gR(\dot{\boldsymbol{\Psi}})$ so that $R(\boldsymbol{a}) = R(\dot{\boldsymbol{\Psi}})$. For $\boldsymbol{a} = \boldsymbol{e}_1$, the equation $R(\boldsymbol{e}_1) = R(\dot{\boldsymbol{\Psi}})$ reads

$$\sin\psi\,\dot{\theta} - \sin\theta\cos\psi\,\dot{\phi} = 1, \quad \cos\psi\,\dot{\theta} + \sin\theta\sin\psi\,\dot{\phi} = 0, \quad \dot{\psi} + \cos\theta\,\dot{\phi} = 0,$$

which are easily solved to give explicit expressions of $\dot{\phi}, \dot{\theta}, \dot{\psi}$, and thereby the left-invariant vector field K_1 is determined. The same procedure runs in parallel for $\boldsymbol{a} = \boldsymbol{e}_2$ and $\boldsymbol{a} = \boldsymbol{e}_3$ to determine K_2, K_3, respectively. It then turns out that the left-invariant vector fields $K_a = gR(\boldsymbol{e}_a)$ are put in the form

$$\begin{cases} K_1 = -\dfrac{\cos\psi}{\sin\theta}\dfrac{\partial}{\partial\phi} + \sin\psi\dfrac{\partial}{\partial\theta} + \cot\theta\cos\psi\dfrac{\partial}{\partial\psi}, \\ K_2 = \dfrac{\sin\psi}{\sin\theta}\dfrac{\partial}{\partial\phi} + \cos\psi\dfrac{\partial}{\partial\theta} - \cot\theta\sin\psi\dfrac{\partial}{\partial\psi}, \\ K_3 = \dfrac{\partial}{\partial\psi}. \end{cases} \tag{1.98}$$

As is easily verified, the K_a are dual to Ψ_a:

$$\Psi^a(K_b) = \delta_{ab}, \quad a, b = 1, 2, 3.$$

In the same manner as above, by using the equation $R(\boldsymbol{a}) = R(\dot{\boldsymbol{\Phi}})$ for $\boldsymbol{a} = \boldsymbol{e}_a, a = 1, 2, 3$, we find that the right-invariant vector fields $J_a = R(\boldsymbol{e}_a)g$ are expressed as

$$\begin{cases} J_1 = -\cos\phi\cot\theta\dfrac{\partial}{\partial\phi} - \sin\phi\dfrac{\partial}{\partial\theta} + \dfrac{\cos\phi}{\sin\theta}\dfrac{\partial}{\partial\psi}, \\ J_2 = -\sin\phi\cot\theta\dfrac{\partial}{\partial\phi} + \cos\phi\dfrac{\partial}{\partial\theta} + \dfrac{\sin\phi}{\sin\theta}\dfrac{\partial}{\partial\psi}, \\ J_3 = \dfrac{\partial}{\partial\phi}, \end{cases} \tag{1.99}$$

which are dual to Φ_a:

$$\Phi^a(J_b) = \delta_{ab}, \quad a, b = 1, 2, 3.$$

In a dual manner to (1.96), the commutation relations among K_a and among J_a are given by

$$[K_a, K_b] = \sum_c \varepsilon_{abc} K_c, \quad [J_a, J_b] = -\sum_c \varepsilon_{abc} J_c, \tag{1.100}$$

respectively. These equations can be verified in a straightforward manner or by using the formula for the differential of a one-form κ, $d\kappa(X, Y) = X(\kappa(Y)) - Y(\kappa(X)) - \kappa([X, Y])$ (see [75]) together with (1.96).

Since $J_a = R(\boldsymbol{e}_a)g$, $K_a = gR(\boldsymbol{e}_a)$, these vector fields are related by

$$\begin{aligned} J_a &= R(\boldsymbol{e}_a)g = gR(g^{-1}\boldsymbol{e}_a) = g\sum g_{ab}R(\boldsymbol{e}_b) \\ &= \sum g_{ab}K_b, \end{aligned} \tag{1.101}$$

where $g = (g_{ab})$ and $g^{-1} = (g_{ab})^T$.

1.6 Spatial Many-Body Systems

We proceed to systems of N particles in the space $\mathbb{R}^3$. Though the geometric setting for spatial many-body systems is an extension of that for planar many-body systems, we need a modification accompanying the extension of the rotation group from $SO(2)$ to $SO(3)$. We denote by $\boldsymbol{x}_\alpha$, $\alpha = 1, 2, \ldots, N$, the position of each particle with mass $m_\alpha > 0$, $\alpha = 1, 2, \ldots, N$. The configuration of the particles is described by $x = (\boldsymbol{x}_1, \boldsymbol{x}_2, \cdots, \boldsymbol{x}_N)$. The totality of the configurations is denoted by

$$X = \{x;\ x = (\boldsymbol{x}_1, \boldsymbol{x}_2, \cdots, \boldsymbol{x}_N),\ \boldsymbol{x}_\alpha \in \mathbb{R}^3\}, \tag{1.102}$$

which is called a configuration space. The X is endowed with a linear space structure by introducing the addition and multiplication operations through

$$(\boldsymbol{x}_1 + \boldsymbol{y}_1, \boldsymbol{x}_2 + \boldsymbol{y}_2, \cdots, \boldsymbol{x}_n + \boldsymbol{y}_N) = (\boldsymbol{x}_1, \boldsymbol{x}_2, \cdots, \boldsymbol{x}_N) + (\boldsymbol{y}_1, \boldsymbol{y}_2, \cdots, \boldsymbol{y}_N),$$

$$\lambda(\boldsymbol{x}_1, \boldsymbol{x}_2, \cdots, \boldsymbol{x}_N) = (\lambda\boldsymbol{x}_1, \lambda\boldsymbol{x}_2, \cdots, \lambda\boldsymbol{x}_N), \quad \lambda \in \mathbb{R}.$$

We can think of the X as the space of $3 \times N$ matrices, since x is viewed as a matrix consisting of N column vectors $\boldsymbol{x}_\alpha$. Here we denote the standard basis vectors $\mathbb{R}^3$ by

$$\boldsymbol{e}_1 = \begin{pmatrix} 1 \\ 0 \\ 0 \end{pmatrix}, \quad \boldsymbol{e}_2 = \begin{pmatrix} 0 \\ 1 \\ 0 \end{pmatrix}, \quad \boldsymbol{e}_3 = \begin{pmatrix} 0 \\ 0 \\ 1 \end{pmatrix}.$$

Then, a basis of X is formed, for example, by

$$(\boldsymbol{e}_k, 0, \cdots, 0), \quad (0, \boldsymbol{e}_k, 0, \cdots, 0), \quad \cdots, \quad (0, \cdots, \boldsymbol{e}_k), \quad k = 1, 2, 3. \tag{1.103}$$

The X is endowed with the mass-weighted inner product (or inner product, for short) through

$$K(x, y) = \sum_{\alpha=1}^{N} m_\alpha \boldsymbol{x}_\alpha \cdot \boldsymbol{y}_\alpha, \quad x, y \in X, \tag{1.104}$$

where the center dot $\cdot$ denotes the standard inner product on $\mathbb{R}^3$. It is easily verified that the K satisfies the following:

1. $K(x, y) = K(y, x)$,
2. $K(\lambda x + \mu y, z) = \lambda K(x, z) + \mu K(y, z), \qquad \lambda, \mu \in \mathbb{R}$,
3. $K(x, x) \geq 0, \qquad K(x, x) = 0 \iff x = 0$.

For any configuration $y = (\boldsymbol{y}_1, \cdots, \boldsymbol{y}_N) \in X$, the center-of-mass vector for y is defined to be

$$\boldsymbol{b} = \left(\sum_{\alpha=1}^{N} m_\alpha \right)^{-1} \sum_{\beta=1}^{N} m_\beta \boldsymbol{y}_\beta.$$

The center-of-mass system is defined to be

$$Q = \{x \in X;\ \sum_{\alpha=1}^{N} m_\alpha \boldsymbol{x}_\alpha = 0\}, \tag{1.105}$$

which is the set of all the configurations with the center-of-mass fixed at the origin of $\mathbb{R}^3$. The Q is a linear subspace of X. By using the center-of-mass vector $\boldsymbol{b}$, any configuration y is decomposed into

$$(\boldsymbol{y}_1, \boldsymbol{y}_2, \cdots, \boldsymbol{y}_N) = (\boldsymbol{x}_1, \boldsymbol{x}_2, \cdots, \boldsymbol{x}_N) + (\boldsymbol{b}, \boldsymbol{b}, \cdots, \boldsymbol{b}), \tag{1.106}$$

where $\boldsymbol{x}_\alpha = \boldsymbol{y}_\alpha - \boldsymbol{b}$. Since $\sum_{\alpha=1}^{N} m_\alpha \boldsymbol{x}_\alpha = 0$, we find that $x = (\boldsymbol{x}_1, \boldsymbol{x}_2, \cdots, \boldsymbol{x}_N) \in Q$ and further that

$$(\boldsymbol{x}_1, \boldsymbol{x}_2, \cdots, \boldsymbol{x}_N) \perp (\boldsymbol{b}, \boldsymbol{b}, \cdots, \boldsymbol{b}).$$

Now we set $b = (\boldsymbol{b}, \cdots, \boldsymbol{b})$. Then, the decomposition (1.106) of y is expressed as $y = x + b$, and proves to be an orthogonal decomposition. On introducing the orthogonal complement to Q by

$$Q^\perp := \{y \in X;\ K(y, x) = 0,\ x \in Q\},$$

the X is decomposed into

$$X = Q \oplus Q^{\perp}. \tag{1.107}$$

The decomposition $y = x + b$ mentioned above is subject to this orthogonal decomposition of X. We note further that

$$Q^{\perp} = \{x \in X \mid x = (\boldsymbol{c}, \boldsymbol{c}, \cdots, \boldsymbol{c}),\ \boldsymbol{c} \in \mathbb{R}^3\}, \tag{1.108}$$

which is easy to prove.

We now seek a basis which fits the decomposition (1.107). We note that the basis (1.103) of X does not fit the decomposition (1.107), since the basis vectors do not belong to the subspace Q or $Q^{\perp}$. A basis of $Q^{\perp}$ is easily found from (1.108), which takes the form $(\boldsymbol{e}_k, \cdots, \boldsymbol{e}_k)$. To find a basis of Q, we take from Q, for example, the vectors $(-m_2\boldsymbol{e}_1, m_1\boldsymbol{e}_1, 0, \cdots, 0)$, $(-m_2\boldsymbol{e}_2, m_1\boldsymbol{e}_2, 0, \cdots, 0)$, $(-m_2\boldsymbol{e}_3, m_1\boldsymbol{e}_3, 0, \cdots, 0)$. We can apply the Gram–Schmidt orthonormalization method to these vectors, and continue the procedure to obtain the following.

Proposition 1.6.1 *The configuration space X for a spatial N-body system has the orthonormal system, with respect to the inner product* (1.104), *given by*

$$\begin{aligned} c_1 &= N_0(\boldsymbol{e}_1, \cdots, \boldsymbol{e}_1), \\ c_2 &= N_0(\boldsymbol{e}_2, \cdots, \boldsymbol{e}_2), \\ c_3 &= N_0(\boldsymbol{e}_3, \cdots, \boldsymbol{e}_3), \end{aligned} \tag{1.109}$$

$$\begin{aligned} f_{3j-2} &= N_j(\overbrace{-m_{j+1}\boldsymbol{e}_1, \cdots, -m_{j+1}\boldsymbol{e}_1}^{j \text{ terms}}, (\sum_{\alpha=1}^{j} m_\alpha)\boldsymbol{e}_1, 0, \cdots, 0), \\ f_{3j-1} &= N_j(\overbrace{-m_{j+1}\boldsymbol{e}_2, \cdots, -m_{j+1}\boldsymbol{e}_2}^{j \text{ terms}}, (\sum_{\alpha=1}^{j} m_\alpha)\boldsymbol{e}_2, 0, \cdots, 0), \\ f_{3j} &= N_j(\overbrace{-m_{j+1}\boldsymbol{e}_3, \cdots, -m_{j+1}\boldsymbol{e}_3}^{j \text{ terms}}, (\sum_{\alpha=1}^{j} m_\alpha)\boldsymbol{e}_3, 0, \cdots, 0), \end{aligned} \tag{1.110}$$

where $j = 1, 2, \cdots, N-1$, and where

$$N_0 = \left(\sum_{\alpha=1}^{N} m_\alpha\right)^{-1/2}, \quad N_j = \left(m_{j+1}\left(\sum_{\alpha=1}^{j} m_\alpha\right)\left(\sum_{\alpha=1}^{j+1} m_\alpha\right)\right)^{-1/2}.$$

Note also that c_k, $k = 1, 2, 3$, and f_ℓ, $\ell = 1, 2, \cdots, 3(N-1)$, form an orthonormal basis of $Q^{\perp}$ and of Q, respectively.

It is straightforward to verify that

$$\begin{aligned} &K(c_i, c_j) = \delta_{ij}, && i, j = 1, 2, 3, \\ &K(c_i, f_k) = 0, && i = 1, 2, 3, \quad k = 1, 2, \ldots, 3(N-1), \\ &K(f_k, f_\ell) = \delta_{k\ell}, && k, \ell = 1, 2, \ldots, 3(N-1). \end{aligned} \tag{1.111}$$

We denote the components of $x \in Q$ with respect to the orthonormal basis f_k by

$$q_k = K(x, f_k), \qquad k = 1, 2, \ldots, 3(N-1). \tag{1.112}$$

Then, the (q_k) serve as the Cartesian coordinates of $Q \cong \mathbb{R}^{3(N-1)}$, which serve to define the Jacobi vectors through

$$\boldsymbol{r}_j := q_{3j-2}\boldsymbol{e}_1 + q_{3j-1}\boldsymbol{e}_2 + q_{3j}\boldsymbol{e}_3, \qquad j = 1, 2, \ldots, N-1. \tag{1.113}$$

It is straightforward to verify that $\boldsymbol{r}_j$ is written out as

$$\begin{aligned} \boldsymbol{r}_j :&= K(x, f_{3j-2})\boldsymbol{e}_1 + K(x, f_{3j-1})\boldsymbol{e}_2 + K(x, f_{3j})\boldsymbol{e}_3 \\ &= N_j \sum_{\alpha=1}^{j} m_\alpha(-m_{j+1})(\boldsymbol{x}_\alpha \cdot \boldsymbol{e}_1)\boldsymbol{e}_1 + m_{j+1}N_j\Big(\sum_{\alpha=1}^{j} m_\alpha\Big)(\boldsymbol{x}_{j+1} \cdot \boldsymbol{e}_1)\boldsymbol{e}_1 \\ &\quad + N_j \sum_{\alpha=1}^{j} m_\alpha(-m_{j+1})(\boldsymbol{x}_\alpha \cdot \boldsymbol{e}_2)\boldsymbol{e}_2 + m_{j+1}N_j\Big(\sum_{\alpha=1}^{j} m_\alpha\Big)(\boldsymbol{x}_{j+1} \cdot \boldsymbol{e}_2)\boldsymbol{e}_2 \\ &\quad + N_j \sum_{\alpha=1}^{j} m_\alpha(-m_{j+1})(\boldsymbol{x}_\alpha \cdot \boldsymbol{e}_3)\boldsymbol{e}_3 + m_{j+1}N_j\Big(\sum_{\alpha=1}^{j} m_\alpha\Big)(\boldsymbol{x}_{j+1} \cdot \boldsymbol{e}_3)\boldsymbol{e}_3 \\ &= -N_j m_{j+1}\Big[\Big(\sum_{\alpha=1}^{j} m_\alpha \boldsymbol{x}_\alpha \cdot \boldsymbol{e}_1\Big)\boldsymbol{e}_1 + \Big(\sum_{\alpha=1}^{j} m_\alpha \boldsymbol{x}_\alpha \cdot \boldsymbol{e}_2\Big)\boldsymbol{e}_2 + \Big(\sum_{\alpha=1}^{j} m_\alpha \boldsymbol{x}_\alpha \cdot \boldsymbol{e}_3\Big)\boldsymbol{e}_3\Big] \\ &\quad + N_j m_{j+1}\Big(\sum_{\alpha=1}^{j} m_\alpha\Big)\Big[(\boldsymbol{x}_{j+1} \cdot \boldsymbol{e}_1)\boldsymbol{e}_1 + (\boldsymbol{x}_{j+1} \cdot \boldsymbol{e}_2)\boldsymbol{e}_2 + (\boldsymbol{x}_{j+1} \cdot \boldsymbol{e}_3)\boldsymbol{e}_3\Big] \\ &= \Big(m_{j+1} \sum_{\alpha=1}^{j} m_\alpha\Big)^{1/2}\Big(\sum_{\alpha=1}^{j+1} m_\alpha\Big)^{-1/2}\Big(\boldsymbol{x}_{j+1} - \Big(\sum_{\alpha=1}^{j} m_\alpha\Big)^{-1} \sum_{\alpha=1}^{j} m_\alpha \boldsymbol{x}_\alpha\Big). \end{aligned}$$

On introducing $M_j = \sum_{\alpha=1}^{j} m_\alpha$, the Jacobi vectors $\boldsymbol{r}_j$ are put in a compact form,

$$\boldsymbol{r}_j = \Big(\frac{1}{M_j} + \frac{1}{m_{j+1}}\Big)^{-1/2}\Big(\boldsymbol{x}_{j+1} - \frac{1}{M_j}\sum_{\alpha=1}^{j} m_\alpha \boldsymbol{x}_\alpha\Big), \quad M_j = \sum_{\alpha=1}^{j} m_\alpha. \tag{1.114}$$

In particular, for $N = 3$, the Jacobi vectors are expressed as

$$\boldsymbol{r}_1 = q_1\boldsymbol{e}_1 + q_2\boldsymbol{e}_2 + q_3\boldsymbol{e}_3 = \sqrt{\frac{m_1 m_2}{m_1 + m_2}}(\boldsymbol{x}_2 - \boldsymbol{x}_1),$$

$$\boldsymbol{r}_2 = q_4\boldsymbol{e}_1 + q_5\boldsymbol{e}_2 + q_6\boldsymbol{e}_3 = \sqrt{\frac{m_3(m_1 + m_2)}{m_1 + m_2 + m_3}}\Big(\boldsymbol{x}_3 - \frac{m_1\boldsymbol{x}_1 + m_2\boldsymbol{x}_2}{m_1 + m_2}\Big).$$

So far we have obtained the isomorphisms, $Q \cong \mathbb{R}^{3(N-1)} \cong \mathbb{R}^{3\times(N-1)}$, given by

$$x = \sum q_k f_k \mapsto (q_k) \mapsto (\boldsymbol{r}_1, \cdots, \boldsymbol{r}_{N-1}),$$

where $\mathbb{R}^{3\times(N-1)}$ denotes the linear space of the real $3 \times (N-1)$ matrices. Thus, the center-of-mass system Q for the spatial N-body system is identified with the space of the $(N-1)$-tuple of the Jacobi vectors $(\boldsymbol{r}_1, \cdots, \boldsymbol{r}_{N-1})$:

$$Q \cong \{(\boldsymbol{r}_1, \cdots, \boldsymbol{r}_{N-1});\ \boldsymbol{r}_j \in \mathbb{R}^3,\ j = 1, \dots, N-1\}. \tag{1.115}$$

For confirmation, we give a procedure to form a configuration $(\boldsymbol{x}_1, \cdots, \boldsymbol{x}_N) \in Q$ from a given set of Jacobi vectors $(\boldsymbol{r}_1, \cdots, \boldsymbol{r}_{N-1})$. To this end, we recall the fact that the Jacobi vector $\boldsymbol{r}_j$ is a constant multiple of the vector whose tail is at the center-of-mass of the particles consisting of the first to the j-th and whose head is at the position of the $(j+1)$-th particle, and further that the $\boldsymbol{r}_{N-1}$ passes the center-of-mass for the whole particle system. Then, the position of each particle is shown to be expressed as

$$\boldsymbol{x}_N = \Big(\frac{1}{M_{N-1}} + \frac{1}{m_N}\Big)^{1/2} \frac{M_{N-1}}{M_{N-1} + m_N}\boldsymbol{r}_{N-1},$$

$$\boldsymbol{x}_{N-1} = \boldsymbol{x}_N - \Big(\frac{1}{M_{N-1}} + \frac{1}{m_N}\Big)^{1/2}\boldsymbol{r}_{N-1} + \Big(\frac{1}{M_{N-2}} + \frac{1}{m_{N-1}}\Big)^{1/2} \frac{M_{N-2}}{M_{N-2} + m_{N-1}}\boldsymbol{r}_{N-2},$$

$$\vdots$$

$$\boldsymbol{x}_2 = \boldsymbol{x}_3 - \Big(\frac{1}{M_2} + \frac{1}{m_3}\Big)^{1/2}\boldsymbol{r}_2 + \Big(\frac{1}{M_1} + \frac{1}{m_2}\Big)^{1/2} \frac{M_1}{M_1 + m_2}\boldsymbol{r}_1,$$

$$\boldsymbol{x}_1 = -\frac{1}{m_1}\sum_{\alpha=2}^{N} m_\alpha \boldsymbol{x}_\alpha.$$

From these equations, the position vectors $\boldsymbol{x}_N, \boldsymbol{x}_{N-1}, \cdots, \boldsymbol{x}_1$ are found to be expressed as linear combinations of the Jacobi vectors $\boldsymbol{r}_{N-1}, \cdots, \boldsymbol{r}_1$. In particular, for $N = 3$, one obtains

$$\boldsymbol{x}_3 = \sqrt{\frac{m_1 + m_2}{(m_1 + m_2 + m_3)m_3}}\boldsymbol{r}_2,$$

$$\boldsymbol{x}_2 = -\sqrt{\frac{m_3}{(m_1 + m_2 + m_3)(m_1 + m_2)}}\boldsymbol{r}_2 + \sqrt{\frac{m_1}{(m_1 + m_2)m_2}}\boldsymbol{r}_1,$$

$$\boldsymbol{x}_1 = -\sqrt{\frac{m_3}{(m_1 + m_2 + m_3)(m_1 + m_2)}}\boldsymbol{r}_2 - \sqrt{\frac{m_2}{m_1(m_1 + m_2)}}\boldsymbol{r}_1.$$

We are now interested in the bulk of the many-body configurations. Among the configurations, there is a special configuration of small bulk. For example, if all the particles lie on a line, the bulk of the configuration is viewed as slender. To describe the bulk or the distribution of particles, we consider the subspace spanned by $\boldsymbol{x}_1, \boldsymbol{x}_2, \cdots, \boldsymbol{x}_N$,

$$W_x = \text{span}\{\boldsymbol{x}_1, \boldsymbol{x}_2, \cdots, \boldsymbol{x}_N\}. \tag{1.116}$$

If all the particles are put together at a point, then one has $\dim W_x = 0$. If the particles are sitting on a line, one has $\dim W_x = 1$. For planar and spatial configurations of particles, we have $\dim W_x = 2$ and $\dim W_x = 3$, respectively. Then, the whole space Q is broken up into subsets, according to $\dim W_x$;

$$Q_k = \{x \in Q;\ \dim W_x = k\}, \quad k = 0, 1, 2, 3. \tag{1.117}$$

In particular, for a reason to be explained soon (see Lemma 1.6.1), we set

$$\dot{Q} = Q_2 \cup Q_3. \tag{1.118}$$

Then, the center-of-mass system Q is broken up into

$$Q = Q_0 \cup Q_1 \cup \dot{Q}. \tag{1.119}$$

We note that this decomposition holds true if we describe the configuration space in terms of the Jacobi vectors (see (1.115)). In fact, if we regard $(\boldsymbol{r}_1, \cdots, \boldsymbol{r}_{N-1})$ as a $3 \times (N-1)$ matrix, then we see that

$$\dim W_x = \text{rank}(\boldsymbol{r}_1, \cdots, \boldsymbol{r}_{N-1}). \tag{1.120}$$

Here we consider the action of the rotation group $SO(3)$ on the center-of-mass system Q. The $SO(3)$ acts on Q in such a manner that $g \in SO(3)$ rotates all the particles simultaneously. Put another way, the action Φ_g of g is defined by

$$\Phi_g(x) = gx = (g\boldsymbol{x}_1, g\boldsymbol{x}_2, \cdots, g\boldsymbol{x}_N), \quad x \in Q. \tag{1.121}$$

Here, we have to note that the Q is invariant under the action of $SO(3)$. In fact, if $\sum m_\alpha \boldsymbol{x}_\alpha = 0$, then $\sum m_\alpha g\boldsymbol{x}_\alpha = 0$. If we consider the system of the Jacobi vectors (1.115) as the configuration space, the $SO(3)$ action is described as

$$(\boldsymbol{r}_1, \boldsymbol{r}_2, \cdots, \boldsymbol{r}_{N-1}) \longmapsto (g\boldsymbol{r}_1, g\boldsymbol{r}_2, \cdots, g\boldsymbol{r}_{N-1}). \tag{1.122}$$

We note further that since $\dim W_{gx} = \dim W_x$, each of the subsets Q_0, Q_1, $\dot{Q}$ is invariant by the action of $SO(3)$ ($g\dot{Q} \subset \dot{Q}$ etc.). The action of $SO(3)$ is then dealt with on each of the subsets Q_0, Q_1, $\dot{Q}$. In particular, as for the $SO(3)$ action on $\dot{Q}$, we can prove the following lemma.

Lemma 1.6.1 *The $SO(3)$ acts freely on $\dot{Q}$, that is, if a point $x \in \dot{Q}$ is fixed by $g \in SO(3)$, then g must be the identity.*

Proof Suppose that there exist $x \in \dot{Q}$ and $g \in SO(3)$ such that $gx = x$. Since $\dim W_x \geq 2$ for $x \in \dot{Q}$, there are at least two linearly independent vectors, which we take as $\boldsymbol{x}_1$, $\boldsymbol{x}_2$ after rearranging the numbering of the position vectors if necessary. From $gx = x$, we have $g\boldsymbol{x}_1 = \boldsymbol{x}_1$, $g\boldsymbol{x}_2 = \boldsymbol{x}_2$, which implies that any vector in the plane spanned by $\boldsymbol{x}_1$ and $\boldsymbol{x}_2$ are left invariant under the action of g. Further, since g is an orthogonal transformation, it preserves the inner product, i.e., for any $\boldsymbol{x}$, $\boldsymbol{y} \in \mathbb{R}^3$, one has $g\boldsymbol{x} \cdot g\boldsymbol{y} = \boldsymbol{x} \cdot \boldsymbol{y}$. It then follows that any vector in the line perpendicular to the plane spanned by $\boldsymbol{x}_1$ and $\boldsymbol{x}_2$ is kept sitting in the same line, if transformed by g. Hence, for any vector $\boldsymbol{n}$ in this line, either of the relations $g\boldsymbol{n} = \pm\boldsymbol{n}$ holds true. Incidentally, the condition $\det g = 1$ requires that $g\boldsymbol{n} = \boldsymbol{n}$. On account of linearity, the transformation g leaves invariant all the vectors in $\mathbb{R}^3$, which means that $g = I$ (the identity). Thus, we have shown that the action of $SO(3)$ is free. This ends the proof.

We now consider the cases other than $\dim W_x \geq 2$. To this end, we introduce the notion of isotropy subgroup. For $x \in Q$, the isotropy subgroup at x is defined to be

$$G_x = \{g \in SO(3);\ gx = x\}. \tag{1.123}$$

Lemma 1.6.2 *According to whether x belongs to Q_0, Q_1 or $\dot{Q}$, the G_x is isomorphic to one of the following;*

$$G_x \cong \begin{cases} SO(3) & \text{if } x \in Q_0, \\ SO(2) & \text{if } x \in Q_1, \\ \{I\} & \text{if } x \in \dot{Q}. \end{cases} \tag{1.124}$$

Proof In terms of isotropy subgroups, Lemma 1.6.1 states that if $x \in \dot{Q}$ then $G_x = \{I\}$. For $x \in Q_0$, one has clearly $G_x \cong SO(3)$. If $x \in Q_1$, all the particles lie on a line. We denote a unit vector on this line by $\boldsymbol{a}$. Then the condition $gx = x$ implies that g is a rotation about the $\boldsymbol{a}$ axis, that is, one has $g = e^{tR(\boldsymbol{a})}$, $t \in \mathbb{R}$. Since there exists an $h \in SO(3)$ such that $\boldsymbol{a} = h\boldsymbol{e}_3$, the rotation about $\boldsymbol{a}$ is expressed as $e^{tR(\boldsymbol{a})} = he^{tR(\boldsymbol{e}_3)}h^{-1}$, which shows that $G_x \cong SO(2)$. This ends the proof.

We now introduce the notion of equivalence relation on Q in such a manner that two configurations are said to be equivalent to each other if they are related by a rotation. In other words, for $x, y \in Q$, the x and y are said to be equivalent if there exists a $g \in SO(3)$ such that $y = gx$. We denote by $x \sim y$ the equivalence of $x, y \in Q$. Then, we can easily verify that (i) $x \sim x$, (ii) $x \sim y \ \Rightarrow y \sim x$, (iii) $x \sim y,\ y \sim z \ \Rightarrow x \sim z$. The quotient space by this equivalence relation, which is denoted by $M := Q/SO(3)$, is called the interior space or the shape space. We denote by π the natural projection from Q to M. Put another way, the π maps $x \in Q$ to its equivalence class $[x]$,

$$\pi : Q \longrightarrow M := Q/SO(3); \quad \pi(x) = [x]. \tag{1.125}$$

We further introduce the notion of the orbit by the group action. The subset of Q defined by

$$\mathcal{O}_x := \{gx \in Q |\, g \in SO(3)\} \tag{1.126}$$

is called the orbit of $SO(3)$ through $x \in Q$. By definition, the $\mathcal{O}_x$ and $[x]$ coincide with each other as subsets. In this sense, the quotient space $Q/SO(3)$ is referred to as the orbit space, which is not a manifold in general. However, if we restrict the center-of-mass system to those $x \in Q$ satisfying $\dim W_x \geq 2$, the quotient space $\dot{M} := \dot{Q}/SO(3)$ becomes a manifold. Here we give an intuitive explanation of this fact. Since $SO(3)$ acts freely on $\dot{Q}$, the equivalence class of $x \in \dot{Q}$ is homeomorphic with $SO(3)$ for any $x \in \dot{Q}$, that is, $\mathcal{O}_x \cong SO(3)$. Hence, the quotient space $\dot{Q}/SO(3)$ is of constant dimension, $\dim \dot{Q}/SO(3) = 3(N-1) - 3 = 3N - 6$. It is a rather difficult problem to identify topologically the shape space. However, we can explicitly find it for a spatial three-body system.

Lemma 1.6.3 *With the condition* $\dim W_x \geq 2$, *the shape space* $\dot{Q}/SO(3)$ *for the center-of-mass system of spatial three bodies is homeomorphic to the upper half space* $\mathbb{R}^3_+$, *where*

$$\mathbb{R}^3_+ = \{(\xi, \eta, \zeta) \in \mathbb{R}^3;\ \zeta > 0\}.$$

Proof From (1.115) with $N = 3$, we view the center-of-mass system in question as $Q = \{(\boldsymbol{r}_1, \boldsymbol{r}_2)|\, \boldsymbol{r}_i \in \mathbb{R}^3,\ i = 1, 2\}$. From (1.120) with $\dim W_x \geq 2$, the $\boldsymbol{r}_1$ and $\boldsymbol{r}_2$ are linearly independent, so that one has $\boldsymbol{r}_1 \times \boldsymbol{r}_2 \neq 0$. In view of this fact, we define a map $\dot{Q} \to \mathbb{R}^3_+$ by

$$(\boldsymbol{r}_1, \boldsymbol{r}_2) \longmapsto (\xi, \eta, \zeta) = (|\boldsymbol{r}_1|^2 - |\boldsymbol{r}_2|^2,\ 2\boldsymbol{r}_1 \cdot \boldsymbol{r}_2,\ 2|\boldsymbol{r}_1 \times \boldsymbol{r}_2|). \tag{1.127}$$

We show that this map gives rise to a map from the restricted shape space $\dot{M} = \dot{Q}/SO(3)$ to $\mathbb{R}^3_+$. According to (1.122), the $SO(3)$ action on $(\boldsymbol{r}_1, \boldsymbol{r}_2)$ is given by $(\boldsymbol{r}_1, \boldsymbol{r}_2) \mapsto (g\boldsymbol{r}_1, g\boldsymbol{r}_2)$. Since the three quantities given in the right-hand side of (1.127) are all invariant under the action of $SO(3)$, they have the same values for equivalent pairs $(\boldsymbol{r}_1, \boldsymbol{r}_2)$, $(\boldsymbol{r}'_1, \boldsymbol{r}'_2) \in \dot{Q}$, so that a map from $\dot{M} = \dot{Q}/SO(3)$ to $\mathbb{R}^3_+$ can be defined by $[(\boldsymbol{r}_1, \boldsymbol{r}_2)] \mapsto (\xi, \eta, \zeta)$. To show that this map is bijective, we refer to the equality

$$(|\boldsymbol{r}_1|^2 - |\boldsymbol{r}_2|^2)^2 + 4(\boldsymbol{r}_1 \cdot \boldsymbol{r}_2)^2 + 4|\boldsymbol{r}_1 \times \boldsymbol{r}_2|^2 = (|\boldsymbol{r}_1|^2 + |\boldsymbol{r}_2|^2)^2. \tag{1.128}$$

For a given $\boldsymbol{a} = \sum_{i=1}^3 a_i \boldsymbol{e}_i \in \mathbb{R}^3_+$ with $a_3 > 0$, we can find a solution to

$$a_1 = |\boldsymbol{r}_1|^2 - |\boldsymbol{r}_2|^2, \quad a_2 = 2\boldsymbol{r}_1 \cdot \boldsymbol{r}_2, \quad a_3 = 2|\boldsymbol{r}_1 \times \boldsymbol{r}_2|,$$

by using (1.128). This is because from $|\boldsymbol{a}| = |\boldsymbol{r}_1|^2 + |\boldsymbol{r}_2|^2$ and $a_1 = |\boldsymbol{r}_1|^2 - |\boldsymbol{r}_2|^2$, the quantities $|\boldsymbol{r}_1|^2$ and $|\boldsymbol{r}_2|^2$ are found to be expressed in terms of a_i, $i = 1, 2, 3$, and then the second and the third equations determine the angle made by $\boldsymbol{r}_1$, $\boldsymbol{r}_2$. Thus, we have found a solution $(\boldsymbol{r}_1, \boldsymbol{r}_2)$. The other solutions are put in the form $(g\boldsymbol{r}_1, g\boldsymbol{r}_2)$, $g \in SO(3)$. In fact, if we have another solution $(\boldsymbol{r}'_1, \boldsymbol{r}'_2)$, the respective magnitudes of $(\boldsymbol{r}'_1, \boldsymbol{r}'_2)$ and the angle made by them are the same as those for $(\boldsymbol{r}_1, \boldsymbol{r}_2)$. Then, one can transfer $\boldsymbol{r}_1$ to $\boldsymbol{r}'_1$ by a rotation g_1 and then by a rotation g_2 about $\boldsymbol{r}'_1$ one can transfer $g_1\boldsymbol{r}_2$ to $\boldsymbol{r}'_2$, that is, $g_2 g_1 \boldsymbol{r}_2 = \boldsymbol{r}'_2$. Thus, we have $(g_2 g_1 \boldsymbol{r}_1, g_2 g_1 \boldsymbol{r}_2) = (\boldsymbol{r}'_1, \boldsymbol{r}'_2)$. This implies that the map of $\dot{M}$ to $\mathbb{R}^3_+$ defined by $[(\boldsymbol{r}_1, \boldsymbol{r}_2)] \mapsto (\xi, \eta, \zeta)$ proves to be a bijection. (If we take into account the topology of the quotient space, we can show that the map $\dot{M} \to \mathbb{R}^3_+$ is actually a homeomorphism.) This ends the proof.

Without restriction to $\dot{Q}$, we consider the whole shape space $Q/SO(3)$ for the three-body system. On account of (1.120), for a three-body system, $\boldsymbol{r}_1$ and $\boldsymbol{r}_2$ are linearly dependent if $\dim W_x = 1$, and hence one has $\boldsymbol{r}_1 \times \boldsymbol{r}_2 = 0$. If we apply the map (1.127) to Q_1, we have

$$(\boldsymbol{r}_1, \boldsymbol{r}_2) \longmapsto (|\boldsymbol{r}_1|^2 - |\boldsymbol{r}_2|^2,\ 2\boldsymbol{r}_1 \cdot \boldsymbol{r}_2,\ 0), \quad (\boldsymbol{r}_1, \boldsymbol{r}_2) \in Q_1.$$

This map gives rise to a bijection from $Q_1/SO(3)$, the quotient of Q_1 by $SO(3)$, to $\dot{\mathbb{R}}^2 := \mathbb{R}^2 - \{0\}$, where $\dot{\mathbb{R}}^2$ denotes the plane $\zeta = 0$ without the origin. The plane $\mathbb{R}^2$ itself forms the boundary of the upper half space $\mathbb{R}^3_+$. In a similar manner, we see that $Q_0/SO(3)$ is nothing but the origin $(\xi, \eta, \zeta) = 0$, which has been deleted from the plane mentioned above. Thus, we have shown the following:

Proposition 1.6.2 *The shape space for the three-body system, $Q/SO(3) = \dot{Q}/SO(3) \cup Q_1/SO(3) \cup Q_0/SO(3)$, is topologically a closed upper half space, which is a disjoint union of three subsets of different dimension:*

$$Q/SO(3) \cong \{(\xi, \eta, \zeta) \in \mathbb{R}^3 : \zeta \geq 0\} = \mathbb{R}^3_+ \cup \dot{\mathbb{R}}^2 \cup \{0\}. \tag{1.129}$$

Shape spaces for spatial N-body systems with $N \geq 4$ are discussed in Appendix 5.3.

1.7 Rotation and Vibration for Spatial Many-Body Systems

In what follows, we treat mainly $\dot{Q}$, and refer to it as the center-of-mass system, and to $\dot{M} = \dot{Q}/SO(3)$ as the shape space accordingly. The inner product defined on $\dot{Q}$ induces the inner product on the tangent space $T_x(\dot{Q})$ at $x \in \dot{Q}$. For tangent vectors $u = (\boldsymbol{u}_1, \boldsymbol{u}_2, \cdots, \boldsymbol{u}_N)$ and $v = (\boldsymbol{v}_1, \boldsymbol{v}_2, \cdots, \boldsymbol{v}_N)$ at $x \in \dot{Q}$, the inner product of them is defined to be

$$K_x(u, v) = \sum_{\alpha=1}^{N} m_\alpha \boldsymbol{u}_\alpha \cdot \boldsymbol{v}_\alpha, \quad u, v \in T_x(\dot{Q}), \tag{1.130}$$

where $\boldsymbol{u}_\alpha \cdot \boldsymbol{v}_\alpha$ denote the standard inner products on $\mathbb{R}^3$ and where $\sum m_\alpha \boldsymbol{u}_\alpha = \sum m_\alpha \boldsymbol{v}_\alpha = 0$ for $u, v \in T_x(\dot{Q})$. The K_x is closely related with the kinetic energy. In fact, for $u = v$, the $K_x(u, u)$ is twice the kinetic energy. The K_x defined by (1.130) is usually described as

$$ds^2 = \sum_{\alpha=1}^{N} m_\alpha d\boldsymbol{x}_\alpha \cdot d\boldsymbol{x}_\alpha, \tag{1.131}$$

which is called a mass-weighted metric (or metric, for short) on $\dot{Q}$.

The first task for us to do is to distinguish rotational and vibrational motions. The rotation of a many-body system is realized by the action of $SO(3)$, which we have already treated. The infinitesimal rotation is called a rotational vector, which is defined as follows: We denote by $R(\boldsymbol{\phi})$ the skew symmetric matrix corresponding to a vector $\boldsymbol{\phi} \in \mathbb{R}^3$. Then the one-parameter group $g(t) = \exp tR(\boldsymbol{\phi})$ determines a curve $g(t)x$ in $\dot{Q}$. Differentiating $g(t)x$ with respect to t at $t = 0$, we obtain the infinitesimal rotation,

$$\left.\frac{d}{dt} g(t)x\right|_{t=0} = R(\boldsymbol{\phi})x = (R(\boldsymbol{\phi})\boldsymbol{x}_1, R(\boldsymbol{\phi})\boldsymbol{x}_2, \cdots, R(\boldsymbol{\phi})\boldsymbol{x}_N). \tag{1.132}$$

In terms of vector fields, the rotational vector is expressed as

$$\sum_{\alpha=1}^{N} R(\boldsymbol{\phi})\boldsymbol{x}_\alpha \cdot \frac{\partial}{\partial \boldsymbol{x}_\alpha} = \boldsymbol{\phi} \cdot \left(\sum_{\alpha=1}^{N} \boldsymbol{x}_\alpha \times \frac{\partial}{\partial \boldsymbol{x}_\alpha} \right), \tag{1.133}$$

where $\partial/\partial \boldsymbol{x}_\alpha$ denotes the gradient operator. The above expression implies that the infinitesimal rotation is closely related with the total angular momentum operator.

We turn to vibrational vectors. A tangent vector $v = (\boldsymbol{v}_1, \boldsymbol{v}_2, \cdots, \boldsymbol{v}_N)$ at x is called a vibrational vector if it is orthogonal to any rotational vector $R(\boldsymbol{\phi})x$ at $x \in \dot{Q}$ with respect to the metric K_x, that is, if v satisfies $K_x(R(\boldsymbol{\phi})x, v) = 0$ for any $\boldsymbol{\phi} \in \mathbb{R}^3$. Written out, this condition is expressed as

$$K_x(R(\boldsymbol{\phi})x, v) = \sum_{\alpha=1}^{N} m_\alpha(\boldsymbol{\phi} \times \boldsymbol{x}_\alpha) \cdot \boldsymbol{v}_\alpha = \boldsymbol{\phi} \cdot \sum_{\alpha=1}^{N} m_\alpha \boldsymbol{x}_\alpha \times \boldsymbol{v}_\alpha = 0,$$

which implies that $v = (\boldsymbol{v}_1, \boldsymbol{v}_2, \cdots, \boldsymbol{v}_N)$ is a vibrational vector if and only if

$$\sum_{\alpha=1}^{N} m_\alpha \boldsymbol{x}_\alpha \times \boldsymbol{v}_\alpha = 0. \tag{1.134}$$

This condition means that the total angular momentum of the N-body system vanishes, which is quite natural as a condition for vibrational vectors from a physical point of view. We note here that the condition $\sum m_\alpha \boldsymbol{v}_\alpha = 0$ is tacitly assumed, which is a condition for v to be a tangent vector to $\dot{Q}$. A vibrational vector v is called also an infinitesimal vibration.

Proposition 1.7.1 *Let $V_{x,\mathrm{rot}}$ and $V_{x,\mathrm{vib}}$ denote the space of rotational and vibrational vectors at $x \in \dot{Q}$, respectively:*

$$V_{x,\mathrm{rot}} = \{R(\boldsymbol{\phi})x;\ \boldsymbol{\phi} \in \mathbb{R}^3\},$$

$$V_{x,\mathrm{vib}} = \{v \in T_x(\dot{Q});\ \sum_{\alpha=1}^{N} m_\alpha \boldsymbol{x}_\alpha \times \boldsymbol{v}_\alpha = 0\} = V_{x,\mathrm{rot}}^{\perp}$$

Then, the tangent space $T_x(\dot{Q})$ is decomposed into the orthogonal direct sum of $V_{x,\mathrm{rot}}$ and $V_{x,\mathrm{vib}}$:

$$T_x(\dot{Q}) = V_{x,\mathrm{rot}} \oplus V_{x,\mathrm{vib}}. \tag{1.135}$$

In terms of differential geometry, the above equation implies that a connection is defined on $\dot{Q}$ and the $V_{x,\mathrm{rot}}$ and $V_{x,\mathrm{vib}}$ are called a vertical and a horizontal subspace, respectively. We call this connection the Guichardet connection [22].

We note also that for $u = (\boldsymbol{u}_1, \cdots, \boldsymbol{u}_N) \in T_{gx}(\dot{Q})$, one has

$$\sum m_\alpha g\boldsymbol{x}_\alpha \times \boldsymbol{u}_\alpha = g \sum m_\alpha \boldsymbol{x}_\alpha \times \boldsymbol{v}_\alpha \quad \text{with} \quad \boldsymbol{v}_\alpha = g^{-1}\boldsymbol{u}_\alpha,$$

which implies that

$$V_{gx,\mathrm{vib}} = gV_{x,\mathrm{vib}}, \quad g \in SO(3). \tag{1.136}$$

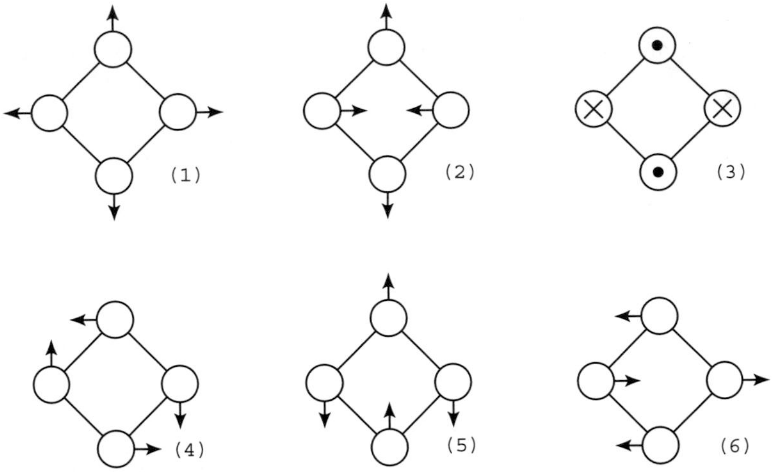

Fig. 1.8 Canonical modes of an X_4 molecule

Here we give a simple example of vibrational vectors. Suppose that there are four particles which are located at $\boldsymbol{x}_1 = \boldsymbol{e}_1, \boldsymbol{x}_2 = \boldsymbol{e}_2, \boldsymbol{x}_3 = -\boldsymbol{e}_1, \boldsymbol{x}_4 = -\boldsymbol{e}_2$. We assume that all particles have an equal mass $m_\alpha = 1$ for simplicity. Then, we can find the following vibrational vectors $e_k, k = 1, \ldots, 6$, satisfying (1.134) with $\boldsymbol{x}_\alpha$ given above:

$$
\begin{aligned}
e_1 &= \tfrac{1}{2}(\boldsymbol{e}_1, \boldsymbol{e}_2, -\boldsymbol{e}_1, -\boldsymbol{e}_2), & e_2 &= \tfrac{1}{2}(-\boldsymbol{e}_1, \boldsymbol{e}_2, \boldsymbol{e}_1, -\boldsymbol{e}_2),\\
e_3 &= \tfrac{1}{2}(-\boldsymbol{e}_3, \boldsymbol{e}_3, -\boldsymbol{e}_3, \boldsymbol{e}_3), & e_4 &= \tfrac{1}{2}(-\boldsymbol{e}_2, -\boldsymbol{e}_1, \boldsymbol{e}_2, \boldsymbol{e}_1),\\
e_5 &= \tfrac{1}{2}(-\boldsymbol{e}_2, \boldsymbol{e}_2, -\boldsymbol{e}_2, \boldsymbol{e}_2), & e_6 &= \tfrac{1}{2}(\boldsymbol{e}_1, -\boldsymbol{e}_1, \boldsymbol{e}_1, -\boldsymbol{e}_1).
\end{aligned}
$$

These vectors form an orthonormal basis of $V_{x,\mathrm{rot}}$ with respect to K_x, and are known also as normal modes for the planar square X_4 molecule (Fig. 1.8).

Any infinitesimal vibration gives rise to an infinitesimal deformation of the shape of the many-body system. This can be seen from (1.125) with Q restricted to $\dot{Q}$. Since the projection $\pi : \dot{Q} \to \dot{M}$ is surjective, the differential of π at $x \in \dot{Q}$

$$
\pi_* : T_x(\dot{Q}) \to T_{\pi(x)}(\dot{M}) \tag{1.137}
$$

is also surjective, whose kernel is shown to be given by

$$
\ker \pi_* = V_{x,\mathrm{rot}}. \tag{1.138}
$$

This and the direct sum decomposition (1.135) of $T_x(\dot{Q})$ are put together to imply that

$$
V_{x,\mathrm{vib}} \cong T_{\pi(x)}(\dot{M}), \tag{1.139}
$$

the proof of which will be given later. The meaning of this equation is that the infinitesimal vibrations at x and the infinitesimal deformation of the shape $\pi(x)$ are in one-to-one correspondence, so that it follows that for any $X \in T_q(\dot{M})$, $q \in \dot{M}$ and $x \in \pi^{-1}(q)$, there exists a unique vibrational vector $u \in V_{x,\mathrm{vib}}$ such that $\pi_*(u) = X$.

While this book does not give a precise definition of a manifold or the definition of the differential of a map either (for those definitions, see [1] and [15], for examples), the differential π_* can be understood in an elementary manner as follows: For any $u \in T_x(\dot{Q})$, we take a curve $x(t)$ which passes x at $t = 0$ with the tangent vector $dx/dt|_{t=0} = u$ at x. Since $\pi(x(t))$ is a curve passing $\pi(x)$ on $\dot{M}$, $d\pi(x(t))/dt|_{t=0}$ is a tangent vector to $\dot{M}$ at $\pi(x)$, which is denoted by $\pi_*(u)$:

$$\pi_*(u) = \frac{d}{dt}\pi(x(t))\bigg|_{t=0} \in T_{\pi(x)}(\dot{M}), \qquad u = \frac{dx}{dt}\bigg|_{t=0}.$$

The map $\pi_* : T_x(\dot{Q}) \to T_{\pi(x)}(\dot{M})$ is thus determined.

With this understanding of π_* in mind, we can verify Eq. (1.138) in the following manner. For $x \in \dot{Q}$, we denote the orbit of $SO(3)$ through the point x by

$$\mathcal{O}_x = \{gx \mid g \in SO(3)\},$$

which consists of all the configurations equivalent to x. Then, from the definition of $V_{x,\mathrm{rot}}$, one has

$$V_{x,\mathrm{rot}} = T_x(\mathcal{O}_x),$$

which implies that the $V_{x,\mathrm{rot}}$ is the tangent space to the orbit $\mathcal{O}_x$. For any $u \in V_{x,\mathrm{rot}}$, there exists a vector $\boldsymbol{\phi} \in \mathbb{R}^3$ such that $u = R(\boldsymbol{\phi})x$ (by the definition of $V_{x,\mathrm{rot}}$). Then, $R(\boldsymbol{\phi})$ is exponentiated to give a one-parameter group, $\exp(tR(\boldsymbol{\phi}))$, whose orbit through x is expressed as $x(t) = \exp(tR(\boldsymbol{\phi}))x$. Since x and $x(t)$ are equivalent with respect to $SO(3)$, we have $\pi(x(t)) = \pi(x)$. Differentiated with respect to t at $t = 0$, this equation provides $\pi_*(R(\boldsymbol{\phi})x) = \pi_*(u) = 0$. This implies that $u \in \ker \pi_*$. Conversely, for any $v \in \ker \pi_*$, there exist a curve $x(t)$ such that $dx(t)/dt|_{t=0} = v$, $x(0) = x$. Differentiation of $\pi(x(t))$ with respect to t results in $d\pi(x(t))/dt|_{t=0} = \pi_*(v) = 0$. If $v \notin V_{x,\mathrm{rot}}$, then v is not tangent to $\mathcal{O}_x$ at $x \in \dot{Q}$. Then for an infinitesimal parameter ε, the $x(\varepsilon)$ does not stay on $\mathcal{O}_x$. Put another way, it leaves $\mathcal{O}_x$. This implies that $x(\varepsilon)$ is not equivalent to x. (If $x(\varepsilon)$ stays in $\mathcal{O}_x$, one has $x(\varepsilon) \in \mathcal{O}_x$, so that $\pi(x(\varepsilon)) = \pi(x)$.) Hence, $\pi(x(\varepsilon)) \neq \pi(x)$, if $\varepsilon \neq 0$. This implies that the curve $\pi(x(t))$ has no stationary point at $\pi(x) \in \dot{M}$. Then, one has $d\pi(x(t))/dt|_{t=0} \neq 0$, which is a contradiction. Therefore, it holds that $v \in V_{x,\mathrm{rot}}$. This ends the proof of $\ker \pi_* = V_{x,\mathrm{rot}}$.

So far we have introduced the connection by using the decomposition (1.135) of the tangent space. There is a dual way to define the connection in terms of an $\mathfrak{so}(3)$-valued differential form. To this end, we need to introduce an inertia tensor. For each $x \in Q$, the inertia tensor $A_x : \mathbb{R}^3 \to \mathbb{R}^3$ is defined through

$$A_x(\boldsymbol{\phi}) = \sum_{\alpha=1}^{N} m_\alpha \boldsymbol{x}_\alpha \times (\boldsymbol{\phi} \times \boldsymbol{x}_\alpha), \quad \boldsymbol{\phi} \in \mathbb{R}^3. \tag{1.140}$$

This definition is an extension of the inertia tensor for a rigid body. For a rigid body, each $\boldsymbol{x}_\alpha$ is relatively fixed, so that one has only to define A_x for a fixed configuration x, but in the present case, the configuration $x \in Q$ can vary without restriction.

Under the condition $\dim W_x \geq 2$, the A_x is shown to be symmetric positive-definite. As the symmetric property is easy to prove, we give here the proof of positive-definiteness only. Since

$$\boldsymbol{\phi} \cdot A_x(\boldsymbol{\phi}) = \sum_{\alpha=1}^{N} m_\alpha (\boldsymbol{\phi} \times \boldsymbol{x}_\alpha) \cdot (\boldsymbol{\phi} \times \boldsymbol{x}_\alpha) \geq 0,$$

the equality of this equation holds if and only if $\boldsymbol{\phi} \times \boldsymbol{x}_\alpha = 0$ for all $\alpha = 1, 2, \ldots, N$. If $\boldsymbol{\phi} \neq 0$, the condition $\boldsymbol{\phi} \times \boldsymbol{x}_\alpha = 0$ implies that there exist $C_\alpha \in \mathbb{R}$ such that $\boldsymbol{x}_\alpha = C_\alpha \boldsymbol{\phi}$. This means that $\dim W_x \leq 1$, which contradicts the assumption $\dim W_x \geq 2$. Then it turns out that $\boldsymbol{\phi} = 0$, as is wanted. Because of the positive-definiteness, the A_x has its inverse, if $x \in \dot{Q}$. In addition, the A_x is shown to transform according to

$$A_{gx}(\boldsymbol{\phi}) = g A_x(g^{-1}\boldsymbol{\phi}), \quad g \in SO(3). \tag{1.141}$$

In fact, we easily verify that

$$\begin{aligned} A_{gx}(\boldsymbol{\phi}) &= \sum_{\alpha=1}^{N} m_\alpha g \boldsymbol{x}_\alpha \times (g g^{-1}\boldsymbol{\phi} \times g \boldsymbol{x}_\alpha) \\ &= g \sum_{\alpha=1}^{N} m_\alpha \boldsymbol{x}_\alpha \times (g^{-1}\boldsymbol{\phi} \times \boldsymbol{x}_\alpha) = g A_x(g^{-1}\boldsymbol{\phi}). \end{aligned}$$

We can put A_x in a matrix form. In fact, the defining equation of $A_x(\boldsymbol{\phi})$ is arranged as

$$A_x(\boldsymbol{\phi}) = \Big(\sum_{\alpha=1}^{N} m_\alpha |\boldsymbol{x}_\alpha|^2 I - \sum_{\alpha=1}^{N} m_\alpha \boldsymbol{x}_\alpha \boldsymbol{x}_\alpha^T \Big) \boldsymbol{\phi}, \tag{1.142}$$

where I denotes the 3×3 identity matrix. Then, the matrix elements of A_x are expressed as

$$\boldsymbol{e}_a \cdot A_x(\boldsymbol{e}_b) = \sum_{\alpha=1}^{N} m_\alpha |\boldsymbol{x}_\alpha|^2 \delta_{ab} - \sum_{\alpha=1}^{N} m_\alpha (\boldsymbol{e}_a \cdot \boldsymbol{x}_\alpha)(\boldsymbol{x}_\alpha \cdot \boldsymbol{e}_b), \quad a, b = 1, 2, 3,$$

where $\boldsymbol{e}_a$ are the standard basis vectors of $\mathbb{R}^3$. See (2.9) and (2.10) for the usual definition of inertia tensor for rigid bodies.

We are now in a position to define the connection form. The connection form ω on $\dot{Q}$ is the $\mathfrak{so}(3)$-valued one-form given by

$$\omega_x = R(A_x^{-1} \sum_{\alpha=1}^{N} m_\alpha \boldsymbol{x}_\alpha \times d\boldsymbol{x}_\alpha). \tag{1.143}$$

By definition, we see that

$$\omega_x(v) = 0 \Leftrightarrow \sum_{\alpha=1}^{N} m_\alpha \boldsymbol{x}_\alpha \times \boldsymbol{v}_\alpha = 0. \tag{1.144}$$

Further, for a rotational vector $R(\boldsymbol{\phi})x$, the ω_x takes the value

$$\omega_x(R(\boldsymbol{\phi})x) = R(A_x^{-1} \sum m_\alpha \boldsymbol{x}_\alpha \times (\boldsymbol{\phi} \times \boldsymbol{x}_\alpha)) = R(\boldsymbol{\phi}). \tag{1.145}$$

We will show that the ω transforms according to

$$(\Phi_g^* \omega)(v) = \mathrm{Ad}_g\, \omega(v), \quad v \in T_x(\dot{Q}), \tag{1.146}$$

where the superscript asterisk denotes the pull-back and the v is viewed as a vector fields on $\dot{Q}$. Before the proof of (1.146), we give a brief review of the pull-back and the differential map. From the action of $SO(3)$ on $\dot{Q}$, the action of $SO(3)$ on the tangent vector fields on $\dot{Q}$ is induced in the following manner: For $u = (\boldsymbol{u}_1, \boldsymbol{u}_2, \cdots, \boldsymbol{u}_N) \in T_x(\dot{Q})$, we consider a curve $x(t)$ which passes the point $x \in \dot{Q}$ at $t = 0$ with u as its tangent vector, $dx/dt|_{t=0} = u$. By the action of Φ_g, this curve is transformed into $gx(t)$. The tangent vector to this curve at gx is given by $g\, dx/dt|_{t=0}$. This provides the definition of the differential map Φ_{g*} of Φ_g. Namely, $\Phi_{g*} : T_x(\dot{Q}) \to T_{gx}(\dot{Q})$ is given by

$$\Phi_{g_*}(u) = (g\boldsymbol{u}_1, g\boldsymbol{u}_2, \cdots, g\boldsymbol{u}_N).$$

We have here to note that Eq. (1.136) should be expressed as $V_{gx,\mathrm{vib}} = \Phi_{g*} V_{x,\mathrm{vib}}$. The pull-back Φ_g^* of Φ_g is now defined in terms of the differential map through

$$(\Phi_g^* \omega)_x(v) := \omega_{gx}(\Phi_{g_*}(v)), \qquad v \in T_x(\dot{Q}).$$

We are now in a position to prove (1.146). The right-hand side of the above equation is arranged by using $A_{gx} = gA_x g^{-1}$, and rewritten as

$$\begin{aligned}\omega_{gx}(\Phi_{g_*}(v)) &= R(A_{gx}^{-1}\sum m_\alpha g\boldsymbol{x}_\alpha \times g\boldsymbol{v}_\alpha)\\ &= gR(A_x^{-1}\sum m_\alpha \boldsymbol{x}_\alpha \times \boldsymbol{v}_\alpha)g^{-1}\\ &= \mathrm{Ad}_g\omega_x(v).\end{aligned}$$

Proposition 1.7.2 *The connection form ω is defined on the center-of-mass system $\dot{Q}$ by* (1.143)*. The ω has the properties described as* (1.144) *and* (1.145)*, and transforms according to* (1.146) *under the action of $SO(3)$.*

In order to see a further meaning of the connection form, we give the decomposition of a tangent vector $v \in T_x(\dot{Q})$ according to the direct sum decomposition (1.135) by using the connection form. Since $\omega_x(v) \in \mathfrak{so}(3)$, the quantity $\omega_x(v)x$ is a rotational vector (see (1.132)). We show that $\omega_x(v)x$ is exactly the rotational component of v. To this end, we introduce the operator P_x by

$$P_x(v) := \omega_x(v)x, \quad P_x(v) = (P_x(v)_1, \cdots, P_x(v)_N).$$

Then, from the definition of $\omega_x(v)$ and that of R, we have

$$P_x(v)_\alpha = (A_x^{-1}\sum_{\beta=1}^{N} m_\beta \boldsymbol{x}_\beta \times \boldsymbol{v}_\beta) \times \boldsymbol{x}_\alpha. \tag{1.147}$$

By using the operator P_x, the tangent vector v is naturally decomposed into $v = P_x(v) + (v - P_x(v))$. Then, a straightforward calculation provides

$$\sum_{\alpha=1}^{N} m_\alpha \boldsymbol{x}_\alpha \times (\boldsymbol{v}_\alpha - P_x(v)_\alpha) = 0, \quad \sum_{\alpha=1}^{N} m_\alpha P_x(v)_\alpha \cdot (\boldsymbol{v}_\alpha - P_x(v)_\alpha) = 0,$$

which implies that the $P_x(v)$ and $v - P_x(v)$ are the rotational and the vibrational components of v, respectively. Thus we find that the P_x is a projection map,

$$P_x : \ T_x(\dot{Q}) \longrightarrow V_{x,\mathrm{rot}}; \quad P_x(v) = \omega_x(v)x.$$

According to the orthogonal decomposition $v = P_x(v) + (v - P_x(v))$, the metric on $T_x(\dot{Q})$ is also decomposed into

$$K_x(v, v) = K_x(P_x(v), P_x(v)) + K_x(v - P_x(v), v - P_x(v)). \tag{1.148}$$

The decomposition (1.148) of the metric and the vector space isomorphism (1.139) are put together to define a Riemannian metric on $\dot{M}$. On account of (1.139), for $X, Y \in T_p(\dot{M})$, there exist vectors $u, v \in V_{x,\mathrm{vib}} \subset T_x(\dot{Q})$ with $\pi(x) = p$ such that $\pi_* u = X$, $\pi_* v = Y$. Then, a Riemannian metric K'_p is defined on $\dot{M}$ through

$$K'_p(X, Y) = K_x(u, v), \quad u, v \in V_{x,\mathrm{vib}}. \tag{1.149}$$

In fact, since K_x is invariant under the action of $SO(3)$, that is $(\Phi_g^* K)_x = K_x$ or

$$K_{gx}(gu, gv) = K_x(u, v), \quad u, v \in T_x(\dot{Q}),$$

and since the invariance $V_{gx,\mathrm{vib}} = \Phi_{g*} V_{x,\mathrm{vib}}$ holds, the right-hand side of (1.149) depends on $\pi(x) \in \dot{M}$ only.

So far we have described the geometric quantities such as the metric, the inertia tensor, and the connection form in terms of position vectors. In the rest of this section, we describe those quantities in terms of Jacobi vectors. We will show that the metric, the inertia tensor, and the connection form can be expressed in terms of the Jacobi vectors (1.114) as follows:

$$ds^2 = \sum_{k=1}^{N-1} d\boldsymbol{r}_k \cdot d\boldsymbol{r}_k, \tag{1.150}$$

$$A_x(\boldsymbol{\phi}) = \sum_{k=1}^{N-1} \boldsymbol{r}_k \times (\boldsymbol{\phi} \times \boldsymbol{r}_k), \tag{1.151}$$

$$\omega = R\big(A_x^{-1} \sum_{k=1}^{N-1} \boldsymbol{r}_k \times d\boldsymbol{r}_k\big). \tag{1.152}$$

We prove these equations by induction with respect to the number $N(\geq 2)$ of particles with attention to the fact that the center-of-mass of the N-particle system and that of the $(N+1)$-particle system are different from each other. Hence, we have to take into account the shift of the center-of-mass when we apply the assumption of induction. As a matter of fact, since the center-of-mass of the N particles in the $(N+1)$-body system is given by $\boldsymbol{X}_N = \sum_{\alpha=1}^{N} m_\alpha \boldsymbol{x}_\alpha / M_N$ with $M_N = \sum_{\alpha=1}^{N} m_\alpha$, we have to translate the position of each particle by $\boldsymbol{x}_\alpha \to \boldsymbol{x}_\alpha - \boldsymbol{X}_N$ in order to apply the assumption of induction.

We start with the proof of (1.150). For $N = 2$, one has $m_1 d\boldsymbol{x}_1 \cdot d\boldsymbol{x}_1 + m_2 d\boldsymbol{x}_2 \cdot d\boldsymbol{x}_2 = d\boldsymbol{r}_1 \cdot d\boldsymbol{r}_1$, as is easily seen. We now assume that Eq. (1.150) holds for N, which is put in the form

$$\sum_{\alpha=1}^{N} m_\alpha d(\boldsymbol{x}_\alpha - \boldsymbol{X}_N) \cdot d(\boldsymbol{x}_\alpha - \boldsymbol{X}_N) = \sum_{k=1}^{N-1} d\boldsymbol{r}_k \cdot d\boldsymbol{r}_k.$$

We note here that $\boldsymbol{r}_k$ with $k = 1, \ldots, N-1$ are invariant under the parallel translation $\boldsymbol{x}_\alpha \to \boldsymbol{x}_\alpha - \boldsymbol{X}_N$. Since $\sum_{\alpha=1}^{N} m_\alpha d\boldsymbol{x}_\alpha = M_N d\boldsymbol{X}_N$, the above equation is rewritten as

$$\sum_{\alpha=1}^{N} m_\alpha d\boldsymbol{x}_\alpha \cdot d\boldsymbol{x}_\alpha = M_N d\boldsymbol{X}_N \cdot d\boldsymbol{X}_N + \sum_{k=1}^{N-1} d\boldsymbol{r}_k \cdot d\boldsymbol{r}_k.$$

We proceed to describe the metric for the $(N+1)$-particle system. A straightforward calculation with this equation provides

$$\begin{aligned}\sum_{\alpha=1}^{N+1} m_\alpha d\boldsymbol{x}_\alpha \cdot d\boldsymbol{x}_\alpha &= M_N d\boldsymbol{X}_N \cdot d\boldsymbol{X}_N + \sum_{k=1}^{N-1} d\boldsymbol{r}_k \cdot d\boldsymbol{r}_k + m_{N+1} d\boldsymbol{x}_{N+1} \cdot d\boldsymbol{x}_{N+1} \\ &= \sum_{k=1}^{N-1} d\boldsymbol{r}_k \cdot d\boldsymbol{r}_k + d\boldsymbol{r}_N \cdot d\boldsymbol{r}_N,\end{aligned}$$

where use has been made of

$$\boldsymbol{x}_{N+1} = \left(\frac{M_N}{m_{N+1}(M_N + m_{N+1})}\right)^{1/2} \boldsymbol{r}_N, \quad \boldsymbol{X}_N = -\left(\frac{m_{N+1}}{M_N(M_N + m_{N+1})}\right)^{1/2} \boldsymbol{r}_N. \tag{1.153}$$

Thus, we have shown that Eq. (1.150) holds for $N + 1$ as well.

We turn to the proof of (1.151). For $N = 2$, one has $m_1 \boldsymbol{x}_1 \times (\boldsymbol{\phi} \times \boldsymbol{x}_1) + m_2 \boldsymbol{x}_2 \times (\boldsymbol{\phi} \times \boldsymbol{x}_2) = \boldsymbol{r}_1 \times (\boldsymbol{\phi} \times \boldsymbol{r}_1)$, as is easily seen. The assumption of induction in the present case is put in the form

$$\sum_{\alpha=1}^{N} m_\alpha (\boldsymbol{x}_\alpha - \boldsymbol{X}_N) \times (\boldsymbol{\phi} \times (\boldsymbol{x}_\alpha - \boldsymbol{X}_N)) = \sum_{k=1}^{N-1} \boldsymbol{r}_k \times (\boldsymbol{\phi} \times \boldsymbol{r}_k).$$

Expanding this equation, we obtain

$$\sum_{\alpha=1}^{N} m_\alpha \boldsymbol{x}_\alpha \times (\boldsymbol{\phi} \times \boldsymbol{x}_\alpha) = M_N \boldsymbol{X}_N \times (\boldsymbol{\phi} \times \boldsymbol{X}_N) + \sum_{k=1}^{N-1} \boldsymbol{r}_k \times (\boldsymbol{\phi} \times \boldsymbol{r}_k).$$

On account of this, the inertia tensor for the $(N+1)$-particle system is expressed and arranged as

$$\sum_{\alpha=1}^{N+1} m_\alpha \boldsymbol{x}_\alpha \times (\boldsymbol{\phi} \times \boldsymbol{x}_\alpha) = M_N \boldsymbol{X}_N \times (\boldsymbol{\phi} \times \boldsymbol{X}_N) + \sum_{k=1}^{N-1} \boldsymbol{r}_k \times (\boldsymbol{\phi} \times \boldsymbol{r}_k)$$
$$+ m_{N+1} \boldsymbol{x}_{N+1} \times (\boldsymbol{\phi} \times \boldsymbol{x}_{N+1})$$
$$= \sum_{k=1}^{N-1} \boldsymbol{r}_k \times (\boldsymbol{\phi} \times \boldsymbol{r}_k) + \boldsymbol{r}_N \times (\boldsymbol{\phi} \times \boldsymbol{r}_N),$$

where use has been made of (1.153). This ends the proof of (1.151).

In conclusion, we prove (1.152). To this end, we have only to show

$$\sum_{\alpha=1}^{N} m_\alpha \boldsymbol{x}_\alpha \times d\boldsymbol{x}_\alpha = \sum_{k=1}^{N-1} \boldsymbol{r}_k \times d\boldsymbol{r}_k. \tag{1.154}$$

This can be proved in the same manner as that used in the proof of (1.150) and (1.151).

1.8 Local Description of Spatial Many-Body Systems

1.8.1 Local Product Structure

In order to make a further study of many-body systems, we need to introduce local coordinate systems on the center-of-mass system $\dot{Q}$. To this end, we use the fact that $\pi : \dot{Q} \to \dot{M}$ is a principal fiber bundle. In place of stating the definition of the principal fiber bundle (see Appendix 5.2), we explain here the local product structure of $\dot{Q}$ as a principal fiber bundle. Let U be an open subset of the shape space $\dot{M}$, that is, the inverse image $\pi^{-1}(U)$ of U is an open subset of $\dot{Q}$. By definition, a given point $q \in U$ determines the shape of the many-body configuration. Then, one can set those particles in the space $\mathbb{R}^3$. A way to do so is for example as follows: For any $x \in \pi^{-1}(q)$, there are at least two linearly independent position vectors, which we denote by $\boldsymbol{x}_1$ and $\boldsymbol{x}_2$ without loss of generality. We can put the $\boldsymbol{x}_1$ on the positive side of the axis $\boldsymbol{e}_1$ by a rotation, keeping the shape invariant, and then the second particle on the $+\boldsymbol{e}_2$ side of the $\boldsymbol{e}_1$–$\boldsymbol{e}_2$ plane, by a rotation about $\boldsymbol{x}_1$. The other particles take their due positions accordingly. If we place the particles in the space

$\mathbb{R}^3$ in this manner, we have a configuration of the many particles for each $q \in U$. This procedure gives us a mapping $\sigma : U \to \dot{Q}$,

$$\sigma(q) = (\boldsymbol{\sigma}_1(q), \boldsymbol{\sigma}_2(q), \cdots, \boldsymbol{\sigma}_N(q)). \tag{1.155}$$

Though the choice of σ is not unique, once σ is given, we can obtain other configurations by rotating $\sigma(q)$. If the subset U is not so large, every configuration x with the shape $\pi(x) \in U$ is realized in this manner:

$$\begin{aligned} x = g\sigma(q), \qquad & g \in SO(3), \\ i.e., \quad (\boldsymbol{x}_1, \cdots, \boldsymbol{x}_N) = (g\boldsymbol{\sigma}_1(q), & \cdots, g\boldsymbol{\sigma}_N(q)). \end{aligned} \tag{1.156}$$

This equation implies that $\pi^{-1}(U)$ has the local product structure

$$\pi^{-1}(U) \cong U \times SO(3).$$

We denote by (q^i), $i = 1, \ldots, 3N - 6$, a local coordinate system on U, where we note that the q^i defined here are different from those defined in (1.112). Natural candidates for local coordinates of $SO(3)$ are the Euler angles. Thus we have obtained a local coordinate system on $\pi^{-1}(U) \cong U \times SO(3)$. If we take the Jacobi vectors to describe the configuration of the center-of-mass system, the lower equation of (1.156) should be replaced by

$$(\boldsymbol{r}_1, \cdots, \boldsymbol{r}_{N-1}) = (g\boldsymbol{\sigma}_1(q), \cdots, g\boldsymbol{\sigma}_{N-1}(q)), \tag{1.157}$$

where $\sigma(q)$ describes a way to place the Jacobi vectors in the space $\mathbb{R}^3$.

It is worth mentioning that the expression (1.156) is closely related to that adopted in a paper by Eckart [17] without reference to the bundle structure of $\dot{Q}$. Let $\boldsymbol{\sigma}_\alpha(q) = \sum_a \boldsymbol{e}_a y_{a\alpha}(q)$. Then, one has, from (1.156),

$$\boldsymbol{x}_\alpha = g \sum_{a=1}^{3} \boldsymbol{e}_a y_{a\alpha}(q) = \sum_a \boldsymbol{\varepsilon}_a y_{a\alpha}(q), \qquad \boldsymbol{\varepsilon}_a := g\boldsymbol{e}_a, \tag{1.158}$$

where $\boldsymbol{\varepsilon}_a, a = 1, 2, 3$, are called the unit vectors along a moving system of axes in [17]. Eckart tried to find a local section suitable for small vibrations of polyatomic molecules, which is precisely realized as the Eckart section in [54] by means of Riemann normal coordinates in a vicinity of an equilibrium point in the shape space and their horizontal lifts to the center-of-mass system. In what follows, we call the standard basis $\{\boldsymbol{e}_a\}$ the space frame, and $\{\boldsymbol{\varepsilon}_a\}$ the rotated frame.

1.8.2 Local Description in the Space Frame

We start with a brief review of rotational vectors and the connection form. Rotational vectors are defined to be infinitesimal generators of the $SO(3)$ action, which are expressed in terms of differential operators (see (1.133)) as

$$\sum_{\alpha=1}^{N}(\boldsymbol{\phi}\times\boldsymbol{x}_\alpha)\cdot\frac{\partial}{\partial\boldsymbol{x}_\alpha}=\boldsymbol{\phi}\cdot\boldsymbol{J},\quad \boldsymbol{\phi}\in\mathbb{R}^3,\quad \boldsymbol{J}=\sum_\alpha \boldsymbol{x}_\alpha\times\frac{\partial}{\partial\boldsymbol{x}_\alpha}. \tag{1.159}$$

The inertia tensor is defined to be

$$A_x(\boldsymbol{v})=\sum_{\alpha=1}^{N} m_\alpha \boldsymbol{x}_\alpha\times(\boldsymbol{v}\times\boldsymbol{x}_\alpha), \tag{1.160}$$

and the connection form ω_x is defined to be

$$\omega_x=R\Big(A_x^{-1}\sum_{\alpha=1}^{N} m_\alpha\boldsymbol{x}_\alpha\times d\boldsymbol{x}_\alpha\Big), \tag{1.161}$$

which satisfies (see (1.145))

$$\omega\big(\boldsymbol{\phi}\cdot\boldsymbol{J}\big)=R(\boldsymbol{\phi}), \tag{1.162}$$

and is subject to the transformation (see (1.146))

$$\omega_{hx}=\mathrm{Ad}_h\omega_x,\quad h\in SO(3). \tag{1.163}$$

We proceed to describe locally-defined connection forms together with rotational and vibrational vectors in local coordinates. Let $\tilde{\omega}^a$ and J_a be components of ω and $\boldsymbol{J}$ with respect to the standard basis vectors $\boldsymbol{e}_a$ (or the space frame), respectively:

$$\omega=\sum R(\boldsymbol{e}_a)\tilde{\omega}^a, \tag{1.164a}$$

$$\boldsymbol{J}=\sum \boldsymbol{e}_a J_a,\quad J_a=\sum_\alpha(\boldsymbol{e}_a\times\boldsymbol{x}_\alpha)\cdot\frac{\partial}{\partial\boldsymbol{x}_\alpha}. \tag{1.164b}$$

We note that from (1.163) the components $\tilde{\omega}^a$ transform according to

$$\tilde{\omega}^a_{hx}=\sum h_{ab}\tilde{\omega}^b_x,\quad h=(h_{ab})\in SO(3). \tag{1.165}$$

On account of the local description $x=g\sigma(q)$, the local expressions of the rotational vectors $(J_a)_x=\frac{d}{dt}e^{tR(\boldsymbol{e}_a)}g\sigma(q)|_{t=0}$ are viewed as right-invariant vector fields already obtained in (1.99). As for the connection form, from (1.162) it follows that $\omega(J_a) = R(\boldsymbol{e}_a)$, which implies that $\tilde{\omega}^a$ are dual to J_a. The forms $\tilde{\omega}^a$ together with dq^i constitute a local basis of the space of one-forms on $\dot{Q}$, which satisfies

$$\tilde{\omega}^a(J_b) = \delta^a_b, \quad dq^i(J_a) = 0, \quad a, b = 1, 2, 3, \quad i = 1, \dots, 3N-6. \tag{1.166}$$

To be precise in notation, we have to use the pull-back $\pi^* dq^i$ for dq^i, but we use dq^i for notational simplicity.

We now set out to obtain a local expression of $\tilde{\omega}^a$. By differentiating $\boldsymbol{x}_\alpha = g\boldsymbol{\sigma}_\alpha(q)$ and arranging the resultant equation, we obtain

$$\sum_\alpha m_\alpha \boldsymbol{x}_\alpha \times d\boldsymbol{x}_\alpha = \sum_\alpha m_\alpha \boldsymbol{x}_\alpha \times dgg^{-1}\boldsymbol{x}_\alpha + \sum_\alpha m_\alpha \boldsymbol{x}_\alpha \times \sum_i \frac{\partial \boldsymbol{x}_\alpha}{\partial q^i} dq^i. \tag{1.167}$$

In view of this, we choose to use the components, Φ^a, of the right-invariant form dgg^{-1} (see (1.93)) and introduce the quantities

$$\gamma_i^a = \left(A_x^{-1} \sum_\alpha m_\alpha \boldsymbol{x}_\alpha \times \frac{\partial \boldsymbol{x}_\alpha}{\partial q^i} \right) \cdot \boldsymbol{e}_a. \tag{1.168}$$

Then, the connection form (1.161) is put in the form

$$\omega = \sum \tilde{\omega}^a R(\boldsymbol{e}_a), \quad \tilde{\omega}^a = \Phi^a + \sum \gamma_i^a dq^i, \tag{1.169}$$

where the local expression of the right-invariant forms Φ^a are given in (1.94).

By using $\tilde{\omega}^a$ and dq^i, we can determine vibrational vector fields X_j through

$$\tilde{\omega}^a(X_j) = 0, \quad dq^i(X_j) = \delta^i_j. \tag{1.170}$$

The X_j and J_a are put together to form a local basis of tangent vector fields on $\dot{Q}$. The local expression of X_j is now found explicitly through (1.170) to be

$$X_j = \frac{\partial}{\partial q^j} - \sum \gamma_j^a J_a, \quad j = 1, 2, \cdots, 3N-6. \tag{1.171}$$

This is a unique vibrational (or horizontal) vector field which is in one-to-one correspondence with $\partial/\partial q^j$ on $U \subset \dot{M}$, and is called the horizontal lift of $\partial/\partial q^j$.

We notice that the transformation rule for $\tilde{\omega}^a$ by g implies that γ_i^a transforms in a similar manner,

$$\gamma_i^a(hx) = \sum h_{ab}\gamma_i^b(x), \quad h = (h_{ab}) \in SO(3). \tag{1.172}$$

For $h = e^{tR(\boldsymbol{e}_c)}$, the above equation is differentiated with respect to t at $t = 0$ to give

$$J_c(\gamma_i^a) = -\sum \varepsilon_{cab}\gamma_i^b, \tag{1.173}$$

where ε_{abc} are the antisymmetric symbols with $\varepsilon_{123} = 1$. In addition, we have to touch upon the transformation property of the components of the inertia tensor. Let $\tilde{A}_{ab}(x) = \boldsymbol{e}_a \cdot A_x(\boldsymbol{e}_b)$. Then, from $A_{hx} = hA_xh^{-1}$ (see (1.141)), we obtain

$$\tilde{A}_{ab}(hx) = \sum_{c,d} h_{ac}\tilde{A}_{cd}(x)h_{bd}, \quad h = (h_{ab}) \in SO(3). \tag{1.174}$$

The infinitesimal transformation of this equation for $h = e^{tR(\boldsymbol{e}_c)}$ proves to be given by

$$J_c(\tilde{A}_{ab}) = \boldsymbol{e}_a \cdot [R(\boldsymbol{e}_c), A_x](\boldsymbol{e}_b) = \sum \varepsilon_{cda}\tilde{A}_{db} + \sum \varepsilon_{cdb}\tilde{A}_{ad}. \tag{1.175}$$

The basis, J_a and X_i, of the tangent vector fields is shown to satisfy the following commutation relations:

$$[J_a, J_b] = -\sum \varepsilon_{abc}J_c, \tag{1.176a}$$

$$[X_i, X_j] = -\sum F_{ij}^c J_c, \tag{1.176b}$$

$$[X_i, J_a] = 0, \tag{1.176c}$$

where F_{ij}^c is defined to be

$$F_{ij}^c := \frac{\partial \gamma_j^c}{\partial q^i} - \frac{\partial \gamma_i^c}{\partial q^j} - \sum_{a,b} \varepsilon_{abc}\gamma_j^a\gamma_j^b, \tag{1.177}$$

and where in the course of calculation, Eq. (1.173) has been effectively used. Equation (1.176b) means that two independent vibrational vector fields, X_i and X_j, are coupled together to give rise to an infinitesimal rotation, if $F_{ij}^c \neq 0$. The quantities F_{ij}^c are called the curvature tensor and will be revisited in Sect. 1.10 (see (1.237)). Put another way, molecular vibrations cannot be separated from rotation, if $F_{ij}^c \neq 0$. Another implication is that the distribution spanned by $\{X_i\}$ is not completely integrable in the sense of Frobenius [75], so that there are no submanifolds to which X_i are tangent.

We proceed to the metric $ds^2 = \sum m_\alpha d\boldsymbol{x}_\alpha \cdot d\boldsymbol{x}_\alpha$, which is decomposed into the sum of vibrational and rotational terms (see (1.148)). We start with the decomposition of the infinitesimal displacement, $dx = (d\boldsymbol{x}_1, \ldots, d\boldsymbol{x}_N)$, by using the local basis, $\tilde{\omega}^a$ and dq^i, of the space of one-forms. A calculation results in

$$dx = \sum_{a=1}^{3} B_a \tilde{\omega}^a + \sum_{i=1}^{3N-6} B_i dq^i, \tag{1.178}$$

or

$$d\boldsymbol{x}_\alpha = \sum_a \boldsymbol{B}_a^\alpha \tilde{\omega}^a + \sum_i \boldsymbol{B}_i^\alpha dq^i, \tag{1.179}$$

where $B_a = (\boldsymbol{B}_a^1, \ldots, \boldsymbol{B}_a^N)$ and $B_i = (\boldsymbol{B}_i^1, \ldots, \boldsymbol{B}_i^N)$ are defined to be

$$\boldsymbol{B}_a^\alpha = J_a \boldsymbol{x}_\alpha = \boldsymbol{e}_a \times \boldsymbol{x}_\alpha, \tag{1.180a}$$

$$\boldsymbol{B}_i^\alpha = X_i \boldsymbol{x}_\alpha = \frac{\partial \boldsymbol{x}_\alpha}{\partial q^i} - A_x^{-1}\Big(\sum_\beta m_\beta \boldsymbol{x}_\beta \times \frac{\partial \boldsymbol{x}_\beta}{\partial q^i}\Big) \times \boldsymbol{x}_\alpha, \tag{1.180b}$$

respectively. The system $\{B_a, B_i\}$ forms a moving frame on the center-of-mass system $\dot{Q}$. From (1.179), the metric $ds^2 = \sum_\alpha m_\alpha d\boldsymbol{x}_\alpha \cdot d\boldsymbol{x}_\alpha$ is written out and arranged as

$$ds^2 = \sum_\alpha \sum_{a,b} m_\alpha \boldsymbol{B}_a^\alpha \cdot \boldsymbol{B}_b^\alpha \tilde{\omega}^a \tilde{\omega}^b + \sum_\alpha \sum_{i,j} m_\alpha \boldsymbol{B}_i^\alpha \cdot \boldsymbol{B}_j^\alpha dq^i dq^j, \tag{1.181}$$

where use has been made of the fact that rotational vectors and vibrational vectors are orthogonal. As for the first term of the right-hand side of the above equation, one easily verifies that

$$\sum_\alpha m_\alpha \boldsymbol{B}_a^\alpha \cdot \boldsymbol{B}_b^\alpha = \boldsymbol{e}_a \cdot A_x(\boldsymbol{e}_b) = \tilde{A}_{ab}. \tag{1.182}$$

The second term of the right-hand side of (1.181) is invariant under the $SO(3)$ action, $\sum_\alpha m_\alpha g\boldsymbol{B}_i^\alpha \cdot g\boldsymbol{B}_j^\alpha = \sum_\alpha m_\alpha \boldsymbol{B}_i^\alpha \cdot \boldsymbol{B}_j^\alpha$, so that it projects to the shape space $\dot{M}$ to define a metric tensor on $\dot{M}$, which we denote by a_{ij},

$$a_{ij} := \sum_\alpha m_\alpha \boldsymbol{B}_i^\alpha \cdot \boldsymbol{B}_j^\alpha. \tag{1.183}$$

Thus, the metric ds^2 on $\dot{Q}$ proves to be expressed as

$$ds^2 = \sum_{a,b} \tilde{A}_{ab} \tilde{\omega}^a \tilde{\omega}^b + \sum_{i,j} a_{ij} dq^i dq^j. \tag{1.184}$$

Riemannian geometry of many-particle systems in terms of the moving frame B_a, B_i is found in [32].

1.8.3 Local Description in the Rotated Frame

So far we have obtained the local description of the inertia tensor, the connection, and the metric in the space frame, in which the right-invariant vector fields and the right-invariant forms on $SO(3)$ have been used. We can obtain another local description in the rotated frame, in which the left-invariant vector fields and the left-invariant forms on $SO(3)$ will be used. We choose here to describe the configuration $x \in \dot{Q}$ in terms of the Jacobi vectors. The inertia tensor, $A_x : \mathbb{R}^3 \to \mathbb{R}^3$, is defined through

$$A_x(\boldsymbol{v}) = \sum_{k=1}^{N-1} \boldsymbol{r}_k \times (\boldsymbol{v} \times \boldsymbol{r}_k), \quad \boldsymbol{v} \in \mathbb{R}^3, \tag{1.185}$$

and the connection form ω is defined for $x \in \dot{Q}$ to be

$$\omega_x = R\Big(A_x^{-1}\big(\sum_{k=1}^{N-1} \boldsymbol{r}_k \times d\boldsymbol{r}_k\big)\Big). \tag{1.186}$$

Needless to say, the transformation properties of A_x and ω_x under the $SO(3)$ action are the same as in the last section.

Local coordinates are introduced through (1.157). On account of $\boldsymbol{r}_k = g\boldsymbol{\sigma}_k(q)$, the connection form ω_x given above with $x = g\sigma(q)$ is written out and arranged as

$$\begin{aligned}
\omega_{g\sigma(q)} &= R\big(A_{g\sigma(q)}^{-1} \sum g\boldsymbol{\sigma}_k(q) \times d(g\boldsymbol{\sigma}_k(q))\big) \\
&= gR\Big(A_{\sigma(q)}^{-1} \sum \big(\boldsymbol{\sigma}_k(q) \times (g^{-1}dg\boldsymbol{\sigma}_k(q) + d\boldsymbol{\sigma}_k(q))\big)\Big)g^{-1} \\
&= gR\Big(A_{\sigma(q)}^{-1} \sum \boldsymbol{\sigma}_k(q) \times (\boldsymbol{\Psi} \times \boldsymbol{\sigma}_k(q))\Big)g^{-1} + gR\Big(A_{\sigma(q)}^{-1} \sum \boldsymbol{\sigma}_k(q) \times d\boldsymbol{\sigma}_k(q)\Big)g^{-1} \\
&= g(g^{-1}dg + \omega_{\sigma(q)})g^{-1},
\end{aligned} \tag{1.187}$$

where use has been made of $R(\boldsymbol{\Psi}) = g^{-1}dg$ and where

$$\omega_{\sigma(q)} := R\Big(A_{\sigma(q)}^{-1}\Big(\sum_{k=1}^{N-1} \boldsymbol{\sigma}_k(q) \times d\boldsymbol{\sigma}_k(q)\Big)\Big). \tag{1.188}$$

We define $\Lambda_i^a(q)$ to be

$$\Lambda_i^a(q) = A_{\sigma(q)}^{-1}\Big(\sum_k \boldsymbol{\sigma}_k \times \frac{\partial \boldsymbol{\sigma}_k}{\partial q^i}\Big)\cdot \boldsymbol{e}_a, \quad a = 1, 2, 3, \ i = 1, \dots, 3N-6, \tag{1.189}$$

and use the components Ψ^a of the left-invariant one-form $g^{-1}dg$ (see (1.93)) to put the connection form ω given by (1.187) in the form

$$\omega_{g\sigma(q)} = \sum_{a=1}^{3} \Theta^a R(g\boldsymbol{e}_a), \quad \Theta^a := \Psi^a + \sum_{i=1}^{3N-6} \Lambda_i^a(q) dq^i, \tag{1.190}$$

where the rotated frame $\{g\boldsymbol{e}_a\}$ has been adopted in contrast to (1.169). We notice here that from the definitions (1.169) and (1.190), $\tilde{\omega}^a$ and Θ^a are related by

$$\tilde{\omega}^a = \sum g_{ab}\Theta^b. \tag{1.191}$$

The horizontal (or vibrational) vector fields Y_j are defined through

$$\Theta^a(Y_j) = 0, \quad dq^i(Y_j) = \delta^i_j, \tag{1.192}$$

and shown to be expressed as

$$Y_j = \frac{\partial}{\partial q^j} - \sum_{a=1}^{3} \Lambda_j^a(q) K_a, \quad j = 1, 2, \cdots, 3N-6, \tag{1.193}$$

where $K_a, a = 1, 2, 3$, denote the left-invariant vector fields on $SO(3)$, which are dual to Ψ^a (see (1.98) and (1.95)). Like X_j given in (1.171), the Y_j is called also the horizontal lift of $\partial/\partial q^j$, where X_j and Y_j are equal to each other in spite of their different expressions. The dq^i, Θ^a and the Y_j, K_a form local basis of one-forms and of vector fields on $\pi^{-1}(U) \cong U \times SO(3)$, respectively. They are dual to each other. The commutation relations among Y_j are shown to be given by

$$[Y_j, Y_j] = -\sum_c \kappa_{ij}^c K_c, \quad \kappa_{ij}^c := \frac{\partial \Lambda_j^c}{\partial q^i} - \frac{\partial \Lambda_i^c}{\partial q^j} - \sum \varepsilon_{abc} \Lambda_i^a \Lambda_j^b, \tag{1.194}$$

where κ_{ij}^c are called also the curvature tensor and related to F_{ij}^c given in (1.177), like (1.191), by

$$F_{ij}^a = \sum g_{ab}\kappa_{ij}^b. \tag{1.195}$$

Like (1.184), we can express the metric $ds^2 = \sum_{k=1}^{N-1} d\boldsymbol{r}_k \cdot d\boldsymbol{r}_k$ in terms of dq^i, Θ^a as

$$ds^2 = \sum_{i,j} a_{ij} dq^i dq^j + \sum_{a,b} A_{ab} \Theta^a \Theta^b, \tag{1.196}$$

where the quantities a_{ij} and A_{ab} are defined to be

$$a_{ij} := ds^2(Y_i, Y_j), \tag{1.197}$$

$$A_{ab} := ds^2\,(K_a, K_b)\,, \tag{1.198}$$

respectively, and where

$$ds^2(Y_i, K_a) = 0, \tag{1.199}$$

as will be verified shortly. We now find explicit expressions of a_{ij} and A_{ab} by writing out the right-hand side of the above equations. Since $g^{-1}dg(K_a) = R(\boldsymbol{e}_a)$, one has $K_a(g) = gR(\boldsymbol{e}_a)$, so that $K_a\boldsymbol{r}_k = K_a(g\boldsymbol{\sigma}_k(q)) = K_a(g)\boldsymbol{\sigma}_k(q) = g(\boldsymbol{e}_a \times \boldsymbol{\sigma}_k(q))$. Hence, we find that the components of the metric tensor on $\dot{M}$ are expressed and arranged as

$$\begin{aligned} a_{ij} &= \sum_k (Y_i\boldsymbol{r}_k) \cdot (Y_j\boldsymbol{r}_k) \\ &= \sum_k \Big(\frac{\partial \boldsymbol{\sigma}_k}{\partial q^i} - \sum_a \Lambda_i^a (\boldsymbol{e}_a \times \boldsymbol{\sigma}_k) \Big) \cdot \Big(\frac{\partial \boldsymbol{\sigma}_k}{\partial q^j} - \sum_b \Lambda_j^b (\boldsymbol{e}_b \times \boldsymbol{\sigma}_k) \Big). \end{aligned} \tag{1.200}$$

The components of the inertial tensor are given by

$$A_{ab} = \sum_k (K_a\boldsymbol{r}_k) \cdot (K_b\boldsymbol{r}_k) = \sum_k (\boldsymbol{e}_a \times \boldsymbol{\sigma}_k) \cdot (\boldsymbol{e}_b \times \boldsymbol{\sigma}_k). \tag{1.201}$$

Equation (1.199) is a consequence of the fact that vibrational vectors are orthogonal to rotational vectors. In fact, since Y_j are vibrational vectors, they are orthogonal to the rotational vectors J_a (see (1.164b)) and since K_a are expressed as linear combinations of J_b (see (1.101)), Y_j are also orthogonal to K_a. Of course, Eq. (1.199) can be verified by calculating $\sum_k d\boldsymbol{r}_k(Y_i) \cdot d\boldsymbol{r}_k(K_a)$ along with the relations $\sum_b A_{ab}\Lambda_i^b = (\sum \boldsymbol{\sigma}_k \times \frac{\partial \boldsymbol{\sigma}_k}{\partial q^i}) \cdot \boldsymbol{e}_a$ and (1.201).

In conclusion of this section, we give the transformation law for local expressions of the connection form and of the inertia tensor. Let $\sigma' : U' \to \dot{Q}$ be another local section with $U' \cap U \neq \emptyset$. Then, there exists an $SO(3)$-valued function $h(q) \in SO(3)$ such that $\sigma'(q) = h(q)\sigma(q)$, $q \in U' \cap U$. From (1.187), it then follows that

$$\omega_{\sigma'(q)} = dhh^{-1} + h\omega_{\sigma(q)}h^{-1}, \tag{1.202}$$

which provides the transformation law

$$\Lambda_i'(q) = \frac{\partial h}{\partial q^i} h^{-1} + h\Lambda_i(q)h^{-1}, \tag{1.203}$$

where

$$\omega_{\sigma'(q)} = \sum_i \Lambda_i'(q)dq^i, \quad \omega_{\sigma(q)} = \sum_i \Lambda_i(q)dq^i. \tag{1.204}$$

Moreover, the transformation law for the inertia tensor $A = (A_{ab})$ is given by

$$A' = hAh^{-1}, \tag{1.205}$$

where

$$A' = \left(A'_{ab}\right), \quad A'_{ab} := \sum_k (\boldsymbol{e}_a \times \boldsymbol{\sigma}'_k) \cdot (\boldsymbol{e}_b \times \boldsymbol{\sigma}'_k). \tag{1.206}$$

1.9 Spatial Three-Body Systems

We introduce local coordinates for a spatial three-body system and express geometric quantities to show that vibrations gives rise to rotations, in particular. In terms of the Jacobi vectors, the center-of-mass system is expressed as $Q = \{(\boldsymbol{r}_1, \boldsymbol{r}_2)\}$. We now define a local section $\sigma : U \to \dot{Q}$ to be

$$\sigma(q) = (\boldsymbol{\sigma}_1(q), \boldsymbol{\sigma}_2(q)), \qquad \boldsymbol{\sigma}_1(q) = q_1\boldsymbol{e}_3, \quad \boldsymbol{\sigma}_2(q) = q_2\boldsymbol{e}_3 + q_3\boldsymbol{e}_1, \tag{1.207}$$

where we have put the first Jacobi vector $\boldsymbol{\sigma}_1$ on the positive side of the $\boldsymbol{e}_3$ axis and the second Jacobi vector $\boldsymbol{\sigma}_2$ on the $+\boldsymbol{e}_1$ side of the $\boldsymbol{e}_3$–$\boldsymbol{e}_1$ plane. We have to restrict the range of (q_1, q_2, q_3) to $q_1 > 0, q_3 > 0$ so that the Jacobi vectors $\boldsymbol{\sigma}_1$ and $\boldsymbol{\sigma}_2$ may be linearly independent. The triple (q_1, q_2, q_3) forms a local coordinate system on an open subset U of the shape space $\dot{M}$. A generic point of $\pi^{-1}(U)$ is described as $g\sigma(q)$ with $g \in SO(3)$ expressed as $g = e^{\phi R(e_3)}e^{\theta R(e_2)}e^{\psi R(e_3)}$ in terms of the Euler angles (ϕ, θ, ψ), so that a local coordinate system on $\pi^{-1}(U)$ is formed by $(\phi, \theta, \psi, q_1, q_2, q_3)$.

For $N - 1 = 2$, Eqs. (1.201) and (1.207) are put together to result in

$$A_{\sigma(q)} = \begin{pmatrix} q_1^2 + q_2^2 & 0 & -q_2q_3 \\ 0 & q_1^2 + q_2^2 + q_3^2 & 0 \\ -q_2q_3 & 0 & q_3^2 \end{pmatrix}. \tag{1.208}$$

The inverse of $A_{\sigma(q)}$ is easily found to be

$$A_{\sigma(q)}^{-1} = \begin{pmatrix} \dfrac{1}{q_1^2} & 0 & \dfrac{q_2}{q_1^2 q_3} \\ 0 & \dfrac{1}{q_1^2+q_2^2+q_3^2} & 0 \\ \dfrac{q_2}{q_1^2 q_3} & 0 & \dfrac{q_1^2+q_2^2}{q_1^2 q_3^2} \end{pmatrix}. \tag{1.209}$$

The connection form at $\sigma(q)$ is then found from (1.188) to be

$$\omega_{\sigma(q)} = R\Big(A_{\sigma(q)}^{-1} \sum_{k=1}^{2} \boldsymbol{\sigma}_k(q) \times d\boldsymbol{\sigma}_k(q)\Big) = \frac{q_2 dq_3 - q_3 dq_2}{q_1^2+q_2^2+q_3^2} R(\boldsymbol{e}_2). \tag{1.210}$$

The connection form at a generic point $x = g\sigma(q)$ is evaluated by using (1.190).

We are now in a position to write down the differential equation for a vibrational motion of the three-body system in terms of the local coordinates. Since a necessary and sufficient condition for a curve $x(t) = g(t)\sigma(q(t))$ to be vibrational is $\omega_{x(t)}(\dot{x}(t)) = 0$, Eqs. (1.187) and (1.210) are put together to provide the differential equation for a vibrational motion in the form

$$g^{-1}\dot{g} + \omega_{\sigma(q)}(\dot{q}) = g^{-1}\frac{dg}{dt} + \frac{q_2\dot{q}_3 - q_3\dot{q}_2}{q_1^2+q_2^2+q_3^2} R(\boldsymbol{e}_2) = 0. \tag{1.211}$$

If $q(t)$ is given, the above differential equation linear in g is easy to integrate at least theoretically. We choose a simple closed curve $q(t)$ given by

$$q_1(t) = c, \quad q_2(t) = a\cos t, \quad q_3(t) = a\sin t, \tag{1.212}$$

where a, b are constants. Then, Eq. (1.211) becomes

$$\frac{dg}{dt} = -gR\Big(\frac{a^2}{c^2+a^2}\boldsymbol{e}_2\Big),$$

which is easily integrated to give

$$g(t) = g(0)\exp\Big(-tR\Big(\frac{a^2}{c^2+a^2}\boldsymbol{e}_2\Big)\Big). \tag{1.213}$$

From this, it follows that the change in g associated with the closed curve $q(t)$ given in (1.212) is evaluated as $g(0)^{-1}g(2\pi) = \exp\Big(-R\Big(\dfrac{2\pi a^2}{c^2+a^2}\boldsymbol{e}_2\Big)\Big)$. This gives the angle, $\dfrac{-2\pi a^2}{c^2+a^2}$, gained by the vibrational motion of the three-body system along with a deformation of shape with the same initial and final shapes. Thus, any rotation angle about the $\boldsymbol{e}_2$-axis can be realized, if the parameters a, c are chosen suitably and if t is replaced by $-t$.

We now inquire why the rotation is made about the $\boldsymbol{e}_2$-axis by vibrational motions of a three-body system in $\mathbb{R}^3$. To this end, we have only to show that the plane spanned by three particles is constant during vibrational motion. Let $\boldsymbol{r}_1$ and $\boldsymbol{r}_2$ be the linearly independent Jacobi vectors, $\boldsymbol{r}_k = g\boldsymbol{\sigma}_k, k = 1, 2$. Then, the unit normal to the plane is given by

$$\boldsymbol{n} = \frac{\boldsymbol{r}_1 \times \boldsymbol{r}_2}{\|\boldsymbol{r}_1 \times \boldsymbol{r}_2\|} = \frac{q_1 q_3 g\boldsymbol{e}_2}{|q_1 q_3|} = \pm g\boldsymbol{e}_2,$$

where the three particles take no collinear configuration, i.e., $q_1 q_3 \neq 0$. By using (1.211), the time-derivative of $\boldsymbol{n}$ is evaluated as

$$\frac{d\boldsymbol{n}}{dt} = \pm \frac{dg}{dt}\boldsymbol{e}_2 = \mp \frac{q_2 \dot{q}_3 - q_3 \dot{q}_2}{q_1^2 + q_2^2 + q_3^2} gR(\boldsymbol{e}_2)\boldsymbol{e}_2 = 0.$$

This implies that $\boldsymbol{n}$ is constant in t, and thereby the vibrational motion takes place on a fixed plane.

So far we have obtained the local expression of the inertia tensor, its inverse, and the connection form for the spatial three-body system. We now find a local expression of the curvature and the metric. From (1.210), the vectors $\boldsymbol{\lambda}_i = \sum \Lambda_i^a \boldsymbol{e}_a$, $i = 1, 2, 3$, associated with the connection form $\omega_{\sigma(q)} = \sum R(\boldsymbol{\lambda}_i) dq^i$ are given by

$$\boldsymbol{\lambda}_1 = \boldsymbol{0}, \qquad \boldsymbol{\lambda}_2 = -\frac{q_3}{q_1^2 + q_2^2 + q_3^2}\boldsymbol{e}_2, \qquad \boldsymbol{\lambda}_3 = \frac{q_2}{q_1^2 + q_2^2 + q_3^2}\boldsymbol{e}_2, \tag{1.214}$$

respectively. From (1.194) and (1.214) with $\Lambda_i = R(\boldsymbol{\lambda}_i)$, the vectors $\boldsymbol{\kappa}_{ij} = \sum \kappa_{ij}^a \boldsymbol{e}_a$ associated with the curvature tensor are calculated as

$$\begin{aligned} &\boldsymbol{\kappa}_{11} = \boldsymbol{\kappa}_{22} = \boldsymbol{\kappa}_{33} = \boldsymbol{0}, \qquad \boldsymbol{\kappa}_{12} = -\boldsymbol{\kappa}_{21} = \frac{2q_1 q_3}{(q_1^2 + q_2^2 + q_3^2)^2}\boldsymbol{e}_2, \\ &\boldsymbol{\kappa}_{23} = -\boldsymbol{\kappa}_{32} = \frac{2q_1^2}{(q_1^2 + q_2^2 + q_3^2)^2}\boldsymbol{e}_2, \qquad \boldsymbol{\kappa}_{31} = -\boldsymbol{\kappa}_{13} = \frac{2q_1 q_2}{(q_1^2 + q_2^2 + q_3^2)^2}\boldsymbol{e}_2. \end{aligned} \tag{1.215}$$

Now that we have verified that $\kappa_{ij}^c \neq 0$, we turn about to look at (1.194), which means that the interference between infinitesimal vibrations results in an infinitesimal rotations. Hence, for the spatial three-body system, infinitesimal vibrations indeed interfere together to give rise to an infinitesimal rotation. In fact, we have already obtained such an example on the level of finite rotations (see (1.213)).

As for the metric tensor on the shape space, by the use of (1.200), we find the expression of the metric tensor and further of its inverse, respectively, as

$$(a_{ij}) = \begin{pmatrix} 1 & 0 & 0 \\ 0 & \dfrac{q_1^2+q_2^2}{q_1^2+q_2^2+q_3^2} & \dfrac{q_2q_3}{q_1^2+q_2^2+q_3^2} \\ 0 & \dfrac{q_2q_3}{q_1^2+q_2^2+q_3^2} & \dfrac{q_1^2+q_3^2}{q_1^2+q_2^2+q_3^2} \end{pmatrix}, \tag{1.216}$$

$$\left(a^{ij}\right) = \begin{pmatrix} 1 & 0 & 0 \\ 0 & \dfrac{q_1^2+q_3^2}{q_1^2} & -\dfrac{q_2q_3}{q_1^2} \\ 0 & -\dfrac{q_2q_3}{q_1^2} & \dfrac{q_1^2+q_2^2}{q_1^2} \end{pmatrix}. \tag{1.217}$$

In the rest of this section, we give another local section, with respect to which the inertia tensor takes a diagonal matrix form. We choose the local section defined to be

$$\boldsymbol{\sigma}_1(q) = \rho\big(\cos\frac{\psi}{2}\cos\frac{\chi}{2}\boldsymbol{e}_1 - \sin\frac{\psi}{2}\sin\frac{\chi}{2}\boldsymbol{e}_2\big), \tag{1.218a}$$

$$\boldsymbol{\sigma}_2(q) = \rho\big(\sin\frac{\psi}{2}\cos\frac{\chi}{2}\boldsymbol{e}_1 + \cos\frac{\psi}{2}\sin\frac{\chi}{2}\boldsymbol{e}_2\big), \tag{1.218b}$$

where $(q^i) = (\rho, \psi, \chi)$ are local coordinates of the shape space $\mathbb{R}^3_+$, as is shown in (1.219). The Jacobi vectors $\boldsymbol{r}_1 = g\boldsymbol{\sigma}_1(q), \boldsymbol{r}_2 = g\boldsymbol{\sigma}_2(q)$ were used in a paper [51] with $\psi/2$ replaced by ψ. From $\boldsymbol{r}_k = g\boldsymbol{\sigma}_k, k = 1, 2$, together with (1.218), we easily verify that the $SO(3)$ invariants given in (1.127) and (1.128) are written out as

$$|\boldsymbol{r}_1|^2 + |\boldsymbol{r}_2|^2 = \rho^2, \tag{1.219a}$$

$$|\boldsymbol{r}_1|^2 - |\boldsymbol{r}_2|^2 = \rho^2\cos\psi\cos\chi, \tag{1.219b}$$

$$2\boldsymbol{r}_1\cdot\boldsymbol{r}_2 = \rho^2\sin\psi\cos\chi, \tag{1.219c}$$

$$2|\boldsymbol{r}_1\times\boldsymbol{r}_2| = \rho^2\sin\chi. \tag{1.219d}$$

where

$$\rho > 0, \quad 0 \le \psi \le 2\pi, \quad 0 < \chi \le \frac{\pi}{2}. \tag{1.220}$$

The characteristic of the local section (1.218) is that the inertia tensor takes a diagonal matrix form with respect to (1.218). In fact, the application of Eqs. (1.201) to (1.218) provides

$$A_{\sigma(q)} = \Big(\sum (\boldsymbol{e}_a \times \boldsymbol{\sigma}_k) \cdot (\boldsymbol{e}_b \times \boldsymbol{\sigma}_k)\Big) = \begin{pmatrix} \rho^2 \sin^2 \frac{\chi}{2} & & \\ & \rho^2 \cos^2 \frac{\chi}{2} & \\ & & \rho^2 \end{pmatrix}. \tag{1.221}$$

We proceed to express the connection form in the local coordinates (ρ, ψ, χ). A straightforward calculation with (1.218) provides

$$\boldsymbol{\sigma}_1 \times d\boldsymbol{\sigma}_1 + \boldsymbol{\sigma}_2 \times d\boldsymbol{\sigma}_2 = -\frac{1}{2}\rho^2 \sin\chi d\psi\, \boldsymbol{e}_3.$$

As $A^{-1}_{\sigma(q)}$ is easily obtained from (1.221), the connection form $\omega_{\sigma(q)}$ proves to be written as

$$\omega_{\sigma(q)} = R\big(A^{-1}_{\sigma(q)} \sum \boldsymbol{\sigma}_k \times d\boldsymbol{\sigma}_k\big) = -\frac{1}{2} \sin\chi d\psi\, R(\boldsymbol{e}_3), \tag{1.222}$$

which shows that the quantities Λ^a_i, $a = 1, 2, 3$, $i = 1, 2, 3$, given in (1.189) are expressed as

$$\Lambda^3_2 = -\frac{1}{2} \sin\chi, \quad \text{the others vanishing.} \tag{1.223}$$

Then, the curvature tensor given in (1.194) is evaluated as

$$\kappa^3_{23} = \frac{1}{2} \cos\chi = -\kappa^3_{32}, \quad \text{the others vanishing.} \tag{1.224}$$

We turn to the metric defined on the shape space. The application of (1.200) to (1.218) and (1.223) provides, after a straightforward but lengthly calculation,

$$(a_{ij}) = \begin{pmatrix} 1 & & \\ & \frac{1}{4}\rho^2 \cos^2\chi & \\ & & \frac{1}{4}\rho^2 \end{pmatrix}. \tag{1.225}$$

In addition, we notice that the quantities $(a^{ij}) = (a_{ij})^{-1}$, $(A^{ab}) = (A_{ab})^{-1}$, $(\det(a_{ij}) \det(A_{ab}))^{1/2}$, and the rotational and vibrational vectors, K_a, Y_j are used in order to describe the kinetic energy operator for the tri-atomic molecule [27].

1.10 Non-separability of Vibration from Rotation

In the last section, we have seen that for the spatial three-body system, vibrational motions give rise to rotations. This means that rotation and vibration cannot be separated for the spatial three-body system. We now ask if this fact holds for the N-body system. To this end, we first consider whether an arbitrary motion can be broken up into vibrational and rotational modes. Let $x(t) = (\boldsymbol{x}_1(t), \cdots, \boldsymbol{x}_N(t))$ be an arbitrary smooth curve in $\dot{Q}$. Our question asks for the decomposition of $x(t)$ into

$$x(t) = g(t)y(t), \quad g(t) \in SO(3), \quad y(t) \in \dot{Q}, \tag{1.226}$$

in such a manner that $y(t)$ may be a vibrational curve. Like (1.187), the connection forms ω_x and ω_y are shown to be related by

$$g^{-1}\omega_x(\dot{x})g = g^{-1}\dot{g} + \omega_y(\dot{y}). \tag{1.227}$$

Since the condition for $y(t)$ to be a vibrational curve is given by $\omega_{y(t)}(\dot{y}(t)) = 0$, the $g(t)$ should be subject to the differential equation

$$\frac{dg}{dt} = \omega_x(\dot{x})g. \tag{1.228}$$

For a given $x(t)$, the present linear differential equation has a unique solution $g(t)$ if an initial condition is given, say, $g(0) = I$, the identity. The vibrational curve is then expressed as $y(t) = g^{-1}(t)x(t)$.

The solution $g(t)$ with $g(0) = I$ has a mechanical implication, as is shown below. Let $\boldsymbol{f}_1(t), \boldsymbol{f}_2(t), \boldsymbol{f}_3(t)$ be the column vectors of $g(t) \in SO(3)$, which serve as a moving frame in $\mathbb{R}^3$. We show that the molecule, the system of particles, does not rotate with respect to the moving frame $\{\boldsymbol{f}_a(t)\}, a = 1, 2, 3$, which is attached to the trajectory of each particle. We denote by $r_\alpha^a(t)$ the components of $\boldsymbol{x}_\alpha(t)$ with respect to $\{\boldsymbol{f}_a(t)\}$,

$$\boldsymbol{x}_\alpha(t) = \sum_a \boldsymbol{f}_a(t) r_\alpha^a(t). \tag{1.229}$$

Let $\boldsymbol{e}_a$ be the standard basis vectors of $\mathbb{R}^3$. Since $g(t)\boldsymbol{e}_a = \boldsymbol{f}_a(t)$ and since $\boldsymbol{x}_\alpha(t) = g(t)\boldsymbol{y}_\alpha(t)$, the above equation is rewritten as

$$\boldsymbol{x}_\alpha(t) = g(t)\boldsymbol{y}_\alpha(t), \quad \boldsymbol{y}_\alpha(t) := \sum \boldsymbol{e}_a r_\alpha^a(t). \tag{1.230}$$

Since $y(t)$ is a vibrational curve with respect to the frame $\{\boldsymbol{e}_a\}$, Eq. (1.229) now means that the motion $x(t)$ is viewed as vibrational with respect to the moving frame $\{\boldsymbol{f}_a(t)\}$. A moving frame with respect to which the molecular motion is vibrational is called an Eckart frame. Thus, we have found that there exists an Eckart frame along any molecular motion, as along as $\dim \mathrm{span}\{\boldsymbol{x}_\alpha(t)\} \geq 2$. The existence of the Eckart frame tempts us to expect that rotation and vibration may be separated. Our next question is whether such an Eckart frame exists independently of the choice of $x(t)$ or not. Suppose that $x(t)$ is a closed curve with $x(0) = x(T)$, $0 \leq t \leq T$. Then, we obtain $x(T) = g(T)y(T) = x(0) = y(0)$. However, as was shown in the preceding section, vibrational motions of the three-body system give rise to rotations, so that $y(T) \neq y(0)$ in general, whereas $\pi(y(T)) = \pi(y(0))$. It will be shown in Appendices 5.5 and 5.6 that vibrational motions of the N-body system also give rise to rotations. Hence, one has $g(T) \neq g(0)$, which means that the Eckart frame defined along a closed curve does not return to the initial one. For this reason, the Eckart frame fails to be defined for closed curves [28].

We study vibrational motions from the viewpoint of total differential equations. We consider the total differential equation

$$\sum_{k=1}^{N-1} \boldsymbol{r}_k \times d\boldsymbol{r}_k = 0. \tag{1.231}$$

If this equation were integrable, then on account of Frobenius' theorem [75], there would be a submanifold on which all motions were vibrational and hence vibrations would be separated from rotations. However, this is not the case. For the spatial three-body system, we have already shown that vibrations indeed give rise to rotations. To deal geometrically with Eq. (1.231), we take a local section σ to express a generic point as $x = g\sigma(q)$. Since Eq. (1.231) is equivalent to $\omega_x = 0$, on account of (1.187), it is equivalently rewritten as

$$g^{-1}dg + \omega_{\sigma(q)} = 0. \tag{1.232}$$

If Eq. (1.232) had a solution, g would be an $SO(3)$-valued function of (q^i). Further, the derivative

$$d^2 g = g(\omega_{\sigma(q)} \wedge \omega_{\sigma(q)} - d\omega_{\sigma(q)}) \tag{1.233}$$

would vanish. Hence, the curvature form

$$\Omega = d\omega - \omega \wedge \omega \tag{1.234}$$

would vanish as well. However, the curvature never vanishes. A proof of this fact will be given in Appendix 5.5 (see Proposition 5.5.5). We will refer to Ω as the curvature form of the Guichardet connection. It then turns out that:

Proposition 1.10.1 *For the N-body configuration space $\dot{Q}$, the following statements are equivalent to one another:*

(1) $\sum_k \boldsymbol{r}_k \times d\boldsymbol{r}_k = 0$ *is not completely integrable.*
(2) *The curvature Ω of the Guichardet connection does not vanish.*
(3) *Vibration cannot be separated from rotation.*
(4) *There does not exist a well-defined moving frame with respect to which a molecule makes vibrations only.*

In the rest of this section, we show that the quantities F_{ij}^c given in (1.177) are exactly the components of the curvature form defined by (1.234). Let the components of Ω be denoted by Ω^a,

$$\Omega = \sum R(\boldsymbol{e}_a)\Omega^a. \tag{1.235}$$

Then, by the definition (1.234) with $\omega = \sum R(\boldsymbol{e}_a)\tilde{\omega}_a$, one verifies that

$$\Omega^c = d\tilde{\omega}^c - \sum \varepsilon_{abc}\tilde{\omega}^a \wedge \tilde{\omega}^b. \tag{1.236}$$

Applying the formula for the one-form θ, $d\theta(X, Y) = X(\theta(Y)) - Y(\theta(X)) - \theta([X, Y])$ (see [75], for example), together with (1.176), we verify that

$$\Omega^c = \sum_{i<j} F_{ij}^c dq^i \wedge dq^j. \tag{1.237}$$

Needless to say, one can derive the above equation by straightforward calculating the right-hand side of Eq. (1.236) with $\tilde{\omega}^c$ given in (1.169). In the course of calculation, the following formula is of great use:

$$d\gamma_j^a = \sum_b J_b(\gamma_j^a)\tilde{\omega}^b + \sum_i X_i(\gamma_j^a)dq^i.$$

In addition, like the connection form ω, the curvature form is subject to the transformation

$$\Omega_{hx} = \mathrm{Ad}_h \Omega_x, \tag{1.238}$$

which is expressed componentwise as

$$\Omega_{hx}^c = \sum h_{ab}\Omega_x^b. \tag{1.239}$$

We remark again that Eq. (1.176b) means that infinitesimal vibrations are coupled together to give rise to an infinitesimal rotation, since $F_{ij}^c \neq 0$ on account of Proposition 5.5.5.

Chapter 2
Mechanics of Many-Body Systems

In this chapter, the equations of motion for a free rigid body are treated on the variational principle, which forms a basis for deriving the equations of motion of spatial many-body systems on the variational principle.

2.1 Equations of Motion for a Free Rigid Body

On the basis of the geometric setting so far for many-body systems, we can formulate Lagrangian and Hamiltonian mechanics. We wish to start with rather simple mechanics for a free rigid body, which will be formulated as a dynamical system on the rotation group $SO(3)$.

If the position of each particle of the N-body system is relatively fixed, or if the shape of the N-body system is fixed, we may consider the N-body system as a rigid body. If a rigid body is a continuum, we have to use continuous parameters to assign the position of each point of the rigid body. With this in mind, we set the center-of-mass of the rigid body fixed at the origin of $\mathbb{R}^3$, and denote the initial position of a point of the body by $\boldsymbol{X}_a$, where a ranges over a finite number of natural numbers if the body is discrete or a is a continuous parameter labeling each point in the body if the body is a continuum. Then, because of the rigidity, possible freedom of motion is only rotation about the center-of-mass, so that the position vector $\boldsymbol{x}_a$ labeled by a is expressed as

$$\boldsymbol{x}_a = g\boldsymbol{X}_a, \quad g \in SO(3).$$

Clearly, we verify that $|\boldsymbol{x}_a - \boldsymbol{x}_b| = |\boldsymbol{X}_a - \boldsymbol{X}_b|$. This implies that the mutual distance of any pair of points of the body is invariant under rotation, which describes the rigidity of the body. From the viewpoint of spatial many-body systems, the shape space reduces to a singleton and the set $(\boldsymbol{X}_a)$ is considered as a section. In addition,

T. Iwai, *Geometry, Mechanics, and Control in Action for the Falling Cat*,
Lecture Notes in Mathematics 2289, https://doi.org/10.1007/978-981-16-0688-5_2

we assume that $\dim \operatorname{span}\{\boldsymbol{X}_a\} \geq 2$, which makes the $SO(3)$ action free on the rigid body.

We denote by m_a a mass attached to the point $\boldsymbol{X}_a$. If a is a continuous parameter, m_a have to be replaced by a continuous distribution of mass. As the velocity vector is written as $\dot{\boldsymbol{x}}_a = \dot{g}\boldsymbol{X}_a$, the momentum vector is expressed as

$$\boldsymbol{p}_a = m_a \dot{g} \boldsymbol{X}_a = m_a g g^{-1} \dot{g} \boldsymbol{X}_a.$$

We point out here that $g^{-1}\dot{g}$ is an anti-symmetric matrix. In fact, if differentiated, $g^{-1}g = I$ yields $\dot{g}^{-1}g + g^{-1}\dot{g} = 0$. Taking into account $\dot{g}^{-1}g = \dot{g}^T g = (g^{-1}\dot{g})^T$, we verify that $g^{-1}\dot{g}$ is anti-symmetric. Since the map $R : \mathbb{R}^3 \to \mathfrak{so}(3)$ is isomorphic, for the anti-symmetric matrix $g^{-1}\dot{g}$, there exists a unique vector $\boldsymbol{\Omega} \in \mathbb{R}^3$ such that

$$R(\boldsymbol{\Omega}) = g^{-1}\dot{g}. \tag{2.1}$$

By using the vector $\boldsymbol{\Omega}$, the momentum $\boldsymbol{p}_a$ is expressed as

$$\boldsymbol{p}_a = m_a g(\boldsymbol{\Omega} \times \boldsymbol{X}_a).$$

We now treat the total angular momentum $\boldsymbol{L}$ of the rigid body. By arranging the defining equation of $\boldsymbol{L}$, we obtain

$$\boldsymbol{L} = \sum_a \boldsymbol{x}_a \times \boldsymbol{p}_a = g \sum_a m_a \boldsymbol{X}_a \times (\boldsymbol{\Omega} \times \boldsymbol{X}_a), \tag{2.2}$$

where the sum should be taken over the whole point of the rigid body. If a is a continuous parameter, we have to take integration over the rigid body in place of the summation. We introduce here the vectors $\boldsymbol{P}_a$ and $\boldsymbol{M}$ by

$$\boldsymbol{P}_a = m_a (\boldsymbol{\Omega} \times \boldsymbol{X}_a),$$

$$\boldsymbol{M} = \sum_a \boldsymbol{X}_a \times \boldsymbol{P}_a = \sum_a m_a \boldsymbol{X}_a \times (\boldsymbol{\Omega} \times \boldsymbol{X}_a), \tag{2.3}$$

respectively. Note that $\boldsymbol{p}_a = g\boldsymbol{P}_a$ and

$$\boldsymbol{L} = g\boldsymbol{M}. \tag{2.4}$$

The $\boldsymbol{L}$ and $\boldsymbol{M}$ are called the total angular momentum in the space frame and that in the body frame, respectively.

In contrast to (2.1), if we use the vector variable $\boldsymbol{\Sigma}$ defined through

$$R(\boldsymbol{\Sigma}) = \dot{g} g^{-1}, \tag{2.5}$$

the angular momentum $\boldsymbol{L}$ is rewritten as

$$\boldsymbol{L} = \sum_a \boldsymbol{x}_a \times \boldsymbol{p}_a = \sum_a m_a \boldsymbol{x}_a \times (\boldsymbol{\Sigma} \times \boldsymbol{x}_a). \tag{2.6}$$

Equations (2.1) and (2.3) and Eqs. (2.5) and (2.6) show that the vector variables $\boldsymbol{\Omega}$ and $\boldsymbol{\Sigma}$ are associated, respectively, with the left- and the right-invariant one-forms (see (1.93)), and related with the angular momentums $\boldsymbol{M}$ and $\boldsymbol{L}$ with respect to the body and the space frames, respectively.

We now recall the definition (1.140) of the inertial tensor. For a rigid body, the inertia tensor $A : \mathbb{R}^3 \to \mathbb{R}^3$ is defined through

$$A\boldsymbol{u} = \sum_a m_a \boldsymbol{X}_a \times (\boldsymbol{u} \times \boldsymbol{X}_a), \qquad \boldsymbol{u} \in \mathbb{R}^3, \tag{2.7}$$

where A is symmetric and positive-definite, if the rigid body spreads in more than one dimension or $\dim \operatorname{span}\{\boldsymbol{X}_a\} \geq 2$. Then, from (2.3) and (2.7), the total angular momentum $\boldsymbol{M}$ is put in the form

$$\boldsymbol{M} = A\boldsymbol{\Omega}. \tag{2.8}$$

For confirmation, we give here the definition of the inertia tensor of a rigid body as a continuum. Let $\boldsymbol{e}_j$, $j = 1, 2, 3$, be an orthonormal frame attached to the rigid body. Then, from (2.7), the matrix elements of A with respect to $\{\boldsymbol{e}_j\}$ are

$$\boldsymbol{e}_i \cdot A\boldsymbol{e}_j = \boldsymbol{e}_i \cdot \boldsymbol{e}_j \sum_a m_a \boldsymbol{X}_a \cdot \boldsymbol{X}_a - \sum_a m_a (\boldsymbol{e}_i \cdot \boldsymbol{X}_a)(\boldsymbol{e}_j \cdot \boldsymbol{X}_a). \tag{2.9}$$

We use $\boldsymbol{r} = \sum_{k=1}^3 x_k \boldsymbol{e}_k$ to assign each point $\boldsymbol{X}_a$ of the rigid body. Then, the summation with weight m_a is transformed into the integration over the volume of the body with respect to $\rho(\boldsymbol{r})dV$, where $\rho(\boldsymbol{r})$ is a mass density distribution and dV denotes the standard volume element. Hence, the matrix elements (2.9) are transformed into

$$\boldsymbol{e}_i \cdot A\boldsymbol{e}_j = \int_V \rho(\boldsymbol{r})(r^2 \delta_{ij} - x_i x_j) dV, \quad r^2 = |\boldsymbol{r}|^2, \quad i, j = 1, 2, 3. \tag{2.10}$$

For a cylinder with radius a of the section and height h, we attach the frame $\boldsymbol{e}_k$ at the center-of-mass in such a manner that $\boldsymbol{e}_3$ is aligned with the (vertical) symmetry axis and $\boldsymbol{e}_1, \boldsymbol{e}_2$ are sitting in the (horizontal) section. It is easy to verify that

$$A = \begin{pmatrix} \frac{1}{12}\mu(3a^2 + \ell^2) & & \\ & \frac{1}{12}\mu(3a^2 + \ell^2) & \\ & & \frac{1}{2}\mu a^2 \end{pmatrix}, \tag{2.11}$$

where μ is the total mass of the cylinder.

Now we are in a position to describe dynamics of the rigid body. The position $g\boldsymbol{X}_a$ of an arbitrary point of the rigid body or its attitude g of the body is determined by $g \in SO(3)$, so that we may take g as a substitute for the position of the rigid body. Then, the velocity of the rigid body is taken as given by $\dot{g}$. It turns out that the state of the rigid body is assigned by $(g, \dot{g})$. We may use $(g, g^{-1}\dot{g})$ in place of $(g, \dot{g})$, since the $(g, \dot{g})$ and the $(g, g^{-1}\dot{g})$ are in one-to-one correspondence. Equivalently, introducing the vector $\boldsymbol{\Omega}$ through $R(\boldsymbol{\Omega}) = g^{-1}\dot{g}$, we may assign the state by the pair $(g, \boldsymbol{\Omega})$ in place of $(g, g^{-1}\dot{g})$. We may further proceed to assign the state by $(g, A\boldsymbol{\Omega}) = (g, \boldsymbol{M})$ with the inertia operator A taken into account, since A is a non-singular matrix. The totality of $(g, \boldsymbol{\Omega})$ or of $(g, \boldsymbol{M})$ forms the space $SO(3) \times \mathbb{R}^3$. Since $(g, \dot{g})$ and $(g, \boldsymbol{\Omega})$ (resp. $(g, \boldsymbol{M})$) are in one-to-one correspondence, the state of the rigid body is represented as a point of $SO(3) \times \mathbb{R}^3$.

Since the pair $(g, \dot{g})$ assigns a point of the tangent bundle $T(SO(3))$ of $SO(3)$, the one-to-one correspondence between $(g, \dot{g})$ and $(g, \boldsymbol{\Omega})$ implies that $T(SO(3))$ can be identified with $SO(3) \times \mathbb{R}^3$. Lagrangian mechanics is described in terms of $(g, \boldsymbol{\Omega})$ and Hamiltonian mechanics in terms of $(g, \boldsymbol{M})$. A reason why Lagrangian and Hamiltonian mechanics are distinguished in this way will be explained in terms of the variational principle in the next section.

If no external force is applied, the total angular momentum $\boldsymbol{L}$ in the space frame is a conserved vector on account of rotational symmetry, which will be verified later in a strict manner. In comparison with this, the total angular momentum $\boldsymbol{M}$ in the body frame is not conserved, but its magnitude is conserved, since $|\boldsymbol{M}| = |\boldsymbol{L}|$. The conservation law for the total angular momentum $\boldsymbol{L}$ in the space frame gives rise to the equation of motion for $\boldsymbol{M}$. In fact, differentiation of $\boldsymbol{L}$ with respect to t results in

$$\frac{d}{dt}\boldsymbol{L} = \frac{d}{dt}(g\boldsymbol{M}) = \frac{dg}{dt}\boldsymbol{M} + g\frac{d}{dt}\boldsymbol{M} = 0.$$

Multiplying this equation by g^{-1} to the left, and using $R(\boldsymbol{\Omega}) = g^{-1}\dot{g},\ \boldsymbol{\Omega} = A^{-1}\boldsymbol{M}$, we obtain

$$\frac{d}{dt}\boldsymbol{M} + A^{-1}\boldsymbol{M} \times \boldsymbol{M} = 0. \tag{2.12}$$

This equation of motion is also expressed, in terms of $\boldsymbol{\Omega}$ for $\boldsymbol{M}$, as

$$\frac{d}{dt}(A\boldsymbol{\Omega}) + \boldsymbol{\Omega} \times A\boldsymbol{\Omega} = 0. \tag{2.13}$$

Both Eqs. (2.12) and (2.13) are known as the Euler equation of motion.

The equation for g is easily given from (2.1) and $\boldsymbol{\Omega} = A^{-1}\boldsymbol{M}$ as follows:

$$\frac{dg}{dt} = gR(A^{-1}\boldsymbol{M}) \quad \text{or} \quad \frac{dg}{dt} = gR(\boldsymbol{\Omega}). \tag{2.14}$$

Thus, we have obtained a complete set of the equations of motion for a free rigid body.

Proposition 2.1.1 *For a free rigid body, the equations of motion in Hamiltonian formalism and in Lagrangian formalism are given on* $SO(3) \times \mathbb{R}^3$ *by* (2.12) *together with the first one of* (2.14) *and by* (2.13) *together with the second one of* (2.14), *respectively.*

We can treat the kinetic energy of the rigid body using the same idea as that implemented for the total angular momentum. A calculation with the arrangement $\dot{\boldsymbol{x}}_a \cdot \dot{\boldsymbol{x}}_a = g^{-1}\dot{g}\boldsymbol{X}_a \cdot g^{-1}\dot{g}\boldsymbol{X}_a$ provides

$$T = \frac{1}{2}\sum_a m_a \dot{\boldsymbol{x}}_a \cdot \dot{\boldsymbol{x}}_a = \frac{1}{2}\boldsymbol{\Omega} \cdot A\boldsymbol{\Omega}. \tag{2.15}$$

Since $\boldsymbol{\Omega} = A^{-1}\boldsymbol{M}$, the above equation is put also in the form

$$T = \frac{1}{2}A^{-1}\boldsymbol{M} \cdot \boldsymbol{M}. \tag{2.16}$$

Solutions to the equations of motion, (2.12) or (2.13), are described in terms of the Jacobi elliptic functions [16]. Once $\boldsymbol{M}$ or $\boldsymbol{\Omega}$ is expressed as functions of t, the equation (2.14) becomes a linear equation for g with $\boldsymbol{M}$ or $\boldsymbol{\Omega}$ being known functions, which can be easily solved. Thus, all the equations of motion are integrated.

To see the behavior of solutions for $\boldsymbol{M}$, we may take another way of expression. From the equation of motion (2.12), we see that the Hamiltonian $H = T$ given in (2.16) is a conserved quantity, $dH/dt = 0$. Since $|\boldsymbol{M}| = |\boldsymbol{L}|$ is known to be conserved, solutions of (2.12) are described as intersections of the sphere $|\boldsymbol{M}| =$ const. with the ellipsoid $H =$ const. The intersections on the ellipsoid are shown in the following figure (Fig. 2.1).

The Euler equation for a free rigid body in $\mathbb{R}^3$ is naturally generalized to that in $\mathbb{R}^n$, which is touched on in Sect. 5.7.

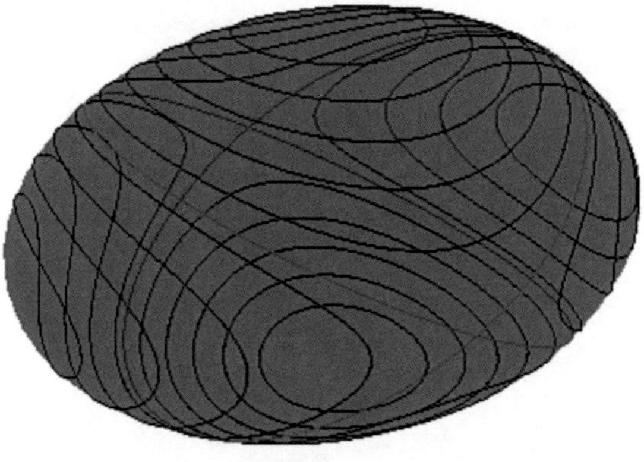

Fig. 2.1 Some solution curves on an ellipsoid on which the Hamiltonian is constant

2.2 Variational Principle for a Free Rigid Body

In this section, we show that the equations of motion are derived on the variational principle both in Lagrangian and Hamiltonian formalisms. The state space for a free rigid body is $SO(3) \times \mathbb{R}^3$.

We start by applying the variational principle in the Lagrangian formalism to obtain the equations of motion (2.13) in terms of the variables $(g, \boldsymbol{\Omega}) \in SO(3) \times \mathbb{R}^3$. The action integral to which the variational principle is applied is given by

$$\int_{t_0}^{t_1} L dt, \qquad L = T = \frac{1}{2} \boldsymbol{\Omega} \cdot A\boldsymbol{\Omega}, \tag{2.17}$$

under the boundary condition for the infinitesimal variation $\delta g(t)$,

$$\delta g(t_j) = 0, \quad j = 0, 1. \tag{2.18}$$

It is to be noted that though the integrand L looks independent of $g \in SO(3)$, the present boundary condition is needed in order to obtain the Euler equation in the Lagrangian formalism. On account of $\langle R(\boldsymbol{a}), R(\boldsymbol{b}) \rangle = \boldsymbol{a} \cdot \boldsymbol{b}$, the infinitesimal variation of the Lagrangian is expressed as

$$\delta L = A\boldsymbol{\Omega} \cdot \delta\boldsymbol{\Omega} = \big\langle R(A\boldsymbol{\Omega}), R(\delta\boldsymbol{\Omega}) \big\rangle.$$

From the definition $R(\boldsymbol{\Omega}) = g^{-1}\dot{g}$, we verify that the variation $R(\delta\boldsymbol{\Omega})$ is arranged as

$$R(\delta\boldsymbol{\Omega}) = \delta(g^{-1}\dot{g}) = [g^{-1}\dot{g}, g^{-1}\delta g] + \frac{d}{dt}(g^{-1}\delta g). \tag{2.19}$$

On using (2.19) and the formula

$$\langle R(\boldsymbol{a}), [R(\boldsymbol{b}), R(\boldsymbol{c})] \rangle = \langle [R(\boldsymbol{a}), R(\boldsymbol{b})], R(\boldsymbol{c}) \rangle, \tag{2.20}$$

which is equivalent to $\boldsymbol{a} \cdot (\boldsymbol{b} \times \boldsymbol{c}) = (\boldsymbol{a} \times \boldsymbol{b}) \cdot \boldsymbol{c}$, the infinitesimal variation of the action integral is arranged, under the boundary condition $\delta g = 0$ at $t = t_0, t_1$, as

$$\begin{aligned} \delta \int_{t_0}^{t_1} L dt &= \int_{t_0}^{t_1} \Big\langle R(A\boldsymbol{\Omega}), [R(\boldsymbol{\Omega}), g^{-1}\delta g] + \frac{d}{dt}(g^{-1}\delta g) \Big\rangle dt \\ &= \int_{t_0}^{t_1} \Big\langle [R(A\boldsymbol{\Omega}), R(\boldsymbol{\Omega})] - \frac{d}{dt} R(A\boldsymbol{\Omega}), g^{-1}\dot{g} \Big\rangle = 0. \end{aligned}$$

Since $g^{-1}\delta g \in \mathfrak{so}(3)$ is arbitrary, the last equality of the above equation implies that

$$\frac{d}{dt} R(A\boldsymbol{\Omega}) + [R(\boldsymbol{\Omega}), R(A\boldsymbol{\Omega})] = 0.$$

By using the formula $[R(\boldsymbol{a}), R(\boldsymbol{b})] = R(\boldsymbol{a} \times \boldsymbol{b})$, we find that

$$\frac{d}{dt}(A\boldsymbol{\Omega}) + \boldsymbol{\Omega} \times A\boldsymbol{\Omega} = 0. \tag{2.21}$$

Thus we have obtained the Euler equation on the variational principle. The equation $g^{-1}\dfrac{dg}{dt} = R(\boldsymbol{\Omega})$, which is a part of the equations of motion, has been used as the definition of $\boldsymbol{\Omega}$ in the course of calculation.

Now that we have obtained the Euler equation on the variational principle, we show that the Euler equation (2.21) is equivalent to the conservation law for the total angular momentum. In fact, differentiating the equation $\boldsymbol{L} = gA\boldsymbol{\Omega}$ with respect to t, we obtain

$$\frac{d\boldsymbol{L}}{dt} = \frac{dg}{dt}A\boldsymbol{\Omega} + g\frac{d}{dt}(A\boldsymbol{\Omega}) = g\Big(\boldsymbol{\Omega} \times A\boldsymbol{\Omega} + \frac{d}{dt}(A\boldsymbol{\Omega})\Big) = 0,$$

which implies the equivalence we wanted to show.

We proceed to show that the equations of motion (2.12) and (2.14) described in terms of $(g, \boldsymbol{M}) \in SO(3) \times \mathbb{R}^3$ are derived as Hamilton's equations of motion with the Hamiltonian (2.16) on the basis of the variational principle in the Hamiltonian formalism. To this end, we introduce the canonical one-form Θ by

$$\Theta = \sum_a \boldsymbol{p}_a \cdot d\boldsymbol{x}_a. \tag{2.22}$$

In view of $g\boldsymbol{X}_a = \boldsymbol{x}_a$ and $g\boldsymbol{P}_a = \boldsymbol{p}_a$, the form Θ is rewritten as

$$\Theta = \sum_a g\boldsymbol{P}_a \cdot dg\boldsymbol{X}_a = \sum_a \boldsymbol{P}_a \cdot g^{-1}dg\boldsymbol{X}_a.$$

On introducing the vector-valued left-invariant one-from $\boldsymbol{\Psi} = \sum \Psi^a \boldsymbol{e}_a$ (see (1.93)), the Θ is put in the form

$$\Theta = \sum_a \boldsymbol{P}_a \cdot (\boldsymbol{\Psi} \times \boldsymbol{X}_a) = \boldsymbol{\Psi} \cdot \sum_a \boldsymbol{X}_a \times \boldsymbol{P}_a = \boldsymbol{\Psi} \cdot \boldsymbol{M}. \tag{2.23}$$

In comparison to this, if we introduce the vector-valued one-form $\boldsymbol{\Phi} = \sum \Phi^a \boldsymbol{e}_a$, the Θ takes the form

$$\Theta = \sum_a \boldsymbol{p}_a \cdot (\boldsymbol{\Phi} \times \boldsymbol{x}_a) = \boldsymbol{\Phi} \cdot \sum_a \boldsymbol{x}_a \times \boldsymbol{p}_a = \boldsymbol{\Phi} \cdot \boldsymbol{L}. \tag{2.24}$$

The one-forms $\boldsymbol{\Psi}$ and $\boldsymbol{\Phi}$ are related by $g\boldsymbol{\Psi} = \boldsymbol{\Phi}$ on account of

$$R(g\boldsymbol{\Psi}) = gR(\boldsymbol{\Psi})g^{-1} = dgg^{-1} = R(\boldsymbol{\Phi}).$$

The integral to which we apply the variational principle in the Hamiltonian formalism is now expressed in terms of $(g, \boldsymbol{M}) \in SO(3) \times \mathbb{R}^3$ as

$$\int_{t_0}^{t_1} \Theta - H dt = \int_{t_0}^{t_1} \boldsymbol{\Psi} \cdot \boldsymbol{M} - H dt, \quad H = T = \frac{1}{2} \boldsymbol{M} \cdot A^{-1} \boldsymbol{M}, \tag{2.25}$$

where the one-form $\boldsymbol{\Psi}$ is to be viewed as evaluated along a curve, that is, $\boldsymbol{\Psi}(\frac{d}{dt})dt = \boldsymbol{\Omega} dt$. We proceed to derive the equation of motion on this setting under the boundary conditions $\delta g(t_i) = 0, i = 1, 2$. The infinitesimal variation of the above integral is expressed as

$$\int_{t_0}^{t_1} \big(\delta\boldsymbol{\Omega} \cdot \boldsymbol{M} + \boldsymbol{\Omega} \cdot \delta\boldsymbol{M} - \delta H\big) dt = 0. \tag{2.26}$$

We first dispose of the term $\delta\boldsymbol{\Omega} \cdot \boldsymbol{M}$. On account of (2.19), the term $\delta\boldsymbol{\Omega} \cdot \boldsymbol{M}$ is rewritten and arranged as

$$\begin{aligned}
\delta\boldsymbol{\Omega} \cdot \boldsymbol{M} &= \big\langle R(\boldsymbol{M}), R(\delta\boldsymbol{\Omega})\big\rangle \\
&= \Big\langle [R(\boldsymbol{M}), g^{-1}\dot{g}] - R\Big(\frac{d\boldsymbol{M}}{dt}\Big), g^{-1}\delta g\Big\rangle + \frac{d}{dt}\langle R(\boldsymbol{M}), g^{-1}\delta g\rangle.
\end{aligned}$$

Integration by parts of the present term under the boundary conditions, $\delta g(t_1) = \delta g(t_2) = 0$, results in

$$\int_{t_0}^{t_1} \delta\boldsymbol{\Omega} \cdot \boldsymbol{M} dt = \int_{t_0}^{t_1} \Big\langle \Big[R(\boldsymbol{M}), g^{-1}\frac{dg}{dt}\Big] - R\Big(\frac{d\boldsymbol{M}}{dt}\Big), g^{-1}\delta g\Big\rangle dt.$$

The integral of the term $\boldsymbol{\Omega} \cdot \delta\boldsymbol{M}$ is rewritten as

$$\int_{t_0}^{t_1} \boldsymbol{\Omega} \cdot \delta\boldsymbol{M} dt = \int_{t_0}^{t_1} \Big\langle g^{-1}\frac{dg}{dt}, R(\delta\boldsymbol{M})\Big\rangle dt.$$

Further, taking $\delta H = A^{-1}\boldsymbol{M} \cdot \delta\boldsymbol{M}$ into account, we finally turn (2.26) into

$$\int_{t_0}^{t_1} \Big\langle \Big[R(\boldsymbol{M}), g^{-1}\frac{dg}{dt}\Big] - R\Big(\frac{d\boldsymbol{M}}{dt}\Big), g^{-1}\delta g\Big\rangle dt + \int_{t_0}^{t_1} \Big\langle g^{-1}\frac{dg}{dt} - R(A^{-1}\boldsymbol{M}), R(\delta\boldsymbol{M})\Big\rangle = 0.$$

Since $g^{-1}\delta g,\ \delta\boldsymbol{M}$ are arbitrary, the above equation implies that

$$\begin{aligned}
&\Big[R(\boldsymbol{M}),\ g^{-1}\frac{dg}{dt}\Big] - R\Big(\frac{d\boldsymbol{M}}{dt}\Big) = 0, \\
&g^{-1}\frac{dg}{dt} - R(A^{-1}\boldsymbol{M}) = 0.
\end{aligned} \tag{2.27}$$

Since

$$\left[R(\boldsymbol{M}),\, g^{-1}\frac{dg}{dt}\right] = [R(\boldsymbol{M}),\, R(A^{-1}\boldsymbol{M})] = R(\boldsymbol{M} \times A^{-1}\boldsymbol{M}),$$

Eq. (2.27) becomes

$$\frac{d\boldsymbol{M}}{dt} = \boldsymbol{M} \times A^{-1}\boldsymbol{M}, \quad \frac{dg}{dt} = gR(A^{-1}\boldsymbol{M}). \tag{2.28}$$

Thus, we have found the Euler equations for a free rigid body on the variational principle in the Hamiltonian formalism.

In closing this section, we describe the Euler equation in terms of the Euler angles. We recall that

$$R(\boldsymbol{\Omega}) = g^{-1}\frac{dg}{dt} = \sum_{k=1}^{3} R(\boldsymbol{e}_k)\Psi^k\Big(\frac{d}{dt}\Big),$$

where $\Psi^k(\frac{d}{dt})$ are evaluated by using $dq^i(d/dt) = \dot{q}^i$ for $(q^i) = (\theta, \phi, \psi)$. Then, it follows from (1.95) that

$$\begin{aligned} \Omega_1 &= \sin\psi\,\dot{\theta} - \sin\theta\cos\psi\,\dot{\phi}, \\ \Omega_2 &= \cos\psi\,\dot{\theta} + \sin\theta\sin\psi\,\dot{\phi}, \\ \Omega_3 &= \dot{\psi} + \cos\theta\,\dot{\phi}. \end{aligned} \tag{2.29}$$

If the inertia tensor is diagonalized in the from $A = \mathrm{diag}(I_1, I_2, I_3)$, the Lagrangian $L = \frac{1}{2}\boldsymbol{\Omega} \cdot A\boldsymbol{\Omega}$ is expressed as

$$\begin{aligned} L =& \frac{1}{2}\Big(I_1(\sin\psi\,\dot{\theta} - \sin\theta\cos\psi\,\dot{\phi})^2 + I_2(\cos\psi\,\dot{\theta} + \sin\theta\sin\psi\,\dot{\phi})^2 \\ &+ I_3(\dot{\psi} + \cos\theta\,\dot{\phi})^2\Big). \end{aligned} \tag{2.30}$$

Lagrange's equations of motion for the free rigid body are obtained as usual on the variational principle, and expressed, in terms of $(\theta, \phi, \psi, \dot{\theta}, \dot{\phi}, \dot{\psi})$, as

$$\frac{d}{dt}\frac{\partial L}{\partial \dot{q}^i} - \frac{\partial L}{\partial q^i} = 0,$$

where $(q^i) = (\theta, \phi, \psi)$ and $(\dot{q}^i) = (\dot{\theta}, \dot{\phi}, \dot{\psi})$.

2.3 Lagrangian Mechanics of Many-Body Systems

In this section, we derive Euler–Lagrange equations for the non-singular configurations (or on $\dot{Q}$) on the variational principle. Further, for a rotationally-invariant Lagrangian, we obtain reduced Euler-Lagrange equations by the use of $SO(3)$ symmetry. In view of the local product structure of $\dot{Q}$, we choose to adopt the local description given in Sect. 1.8.3.

Let U be an open subset of $\dot{M}$ and $(q, g, \dot{q}, \dot{g})$ be local coordinates on the tangent bundle $T(\pi^{-1}(U))$, where $(q, g) \in \pi^{-1}(U)$ and $(\dot{q}, \dot{g}) \in T_{g\sigma(q)}(\pi^{-1}(U))$ with $q = (q^i)$, $\dot{q} = (\dot{q}^i)$. In association with the connection form Θ^a given by (1.190), we introduce an $\mathfrak{so}(3)$-valued variable by

$$\Pi = \xi + \sum_{i=1}^{3N-6} \Lambda_i(q)\dot{q}^i, \tag{2.31}$$

where

$$\xi = g^{-1}\dot{g}, \quad \Lambda_i(q) = \sum_{a=1}^{3} \Lambda_i^a(q) R(\boldsymbol{e}_a). \tag{2.32}$$

We note that the variable Π reduces to $R(\boldsymbol{\Omega})$ given in (2.1) if the configuration of the system is rigid, i.e., $\dot{q}^i = 0$. Put another way, the Π is the extension of the variable $R(\boldsymbol{\Omega})$ for the rigid body to the $\mathfrak{so}(3)$-valued variable for the deformable body. We take $(q, g, \dot{q}, \Pi)$ as local coordinates in $T(\pi^{-1}(U))$. For a given Lagrangian $L(q, g, \dot{q}, \Pi)$, we wish to obtain the Euler–Lagrange equation on the basis of the variational principle through

$$\delta \int_{t_1}^{t_2} L(q, g, \dot{q}, \Pi) dt = 0 \tag{2.33}$$

under the boundary conditions for the infinitesimal variations $\delta q(t)$ and $\delta g(t)$,

$$\delta q(t_k) = 0, \quad \delta g(t_k) = 0, \quad k = 1, 2. \tag{2.34}$$

Before dealing with variations of the Lagrangian L, we have to discuss derivatives of functions on $SO(3) \times \mathfrak{so}(3)$. The functional derivative [57] of L with respect to $\Pi \in \mathfrak{so}(3)$ is defined through

$$\frac{d}{dt} L(\cdots, \Pi + t\zeta)\Big|_{t=0} = \left\langle \frac{\delta L}{\delta \Pi}, \zeta \right\rangle, \quad \zeta \in \mathfrak{so}(3), \tag{2.35}$$

where the inner product on $\mathfrak{so}(3)$ has been adopted (see (1.86)). A straightforward calculation of the left-hand side provides

$$\frac{d}{dt}L(\cdots, \Pi + t\zeta)\Big|_{t=0} = \sum_{i,j} \frac{\partial L}{\partial \Pi_{ij}} \frac{d}{dt}(\Pi + t\zeta)_{ij}\Big|_{t=0}.$$

Correspondingly, the variation of L with respect to $\delta\Pi$ is expressed as

$$\sum_{i,j} \frac{\partial L}{\partial \Pi_{ij}} \delta\Pi_{ij} = \left\langle \frac{\delta L}{\delta \Pi}, \delta\Pi \right\rangle. \tag{2.36}$$

In a similar manner, as a function on $SO(3)$, the derivative of $L(\cdots, e^{t\eta}g, \cdot)$ with respect to t at $t = 0$ is evaluated as

$$\frac{d}{dt}L(\cdots, e^{t\eta}g, \cdot)\Big|_{t=0} = \sum_{i,j} \frac{\partial L}{\partial g_{ij}} \frac{d}{dt}(e^{t\eta}g)_{ij}\Big|_{t=0} = \left\langle \frac{\partial L}{\partial g} g^{-1} - g\left(\frac{\partial L}{\partial g}\right)^T, \eta \right\rangle, \tag{2.37}$$

where $\eta \in \mathfrak{so}(3)$, so that the corresponding variation of L with respect to $g \in SO(3)$ is given by

$$\sum_{i,j} \frac{\partial L}{\partial g_{ij}} \delta g_{ij} = \left\langle \frac{\partial L}{\partial g} g^{-1} - g\left(\frac{\partial L}{\partial g}\right)^T, \delta g g^{-1} \right\rangle. \tag{2.38}$$

In comparison, the derivative of $L(\cdots, ge^{t\zeta}, \cdot)$ with $\zeta \in \mathfrak{so}(3)$ is expressed as

$$\frac{d}{dt}L(\cdots, ge^{t\zeta}, \cdot)\Big|_{t=0} = \sum_{i,j} \frac{\partial L}{\partial g_{ij}} \frac{d}{dt}(ge^{t\zeta})_{ij}\Big|_{t=0} = \left\langle g^{-1}\frac{\partial L}{\partial g} - \left(\frac{\partial L}{\partial g}\right)^T g, \zeta \right\rangle, \tag{2.39}$$

and the corresponding variation of L with respect to $g \in SO(3)$ is given by

$$\sum_{i,j} \frac{\partial L}{\partial g_{ij}} \delta g_{ij} = \left\langle g^{-1}\frac{\partial L}{\partial g} - \left(\frac{\partial L}{\partial g}\right)^T g, g^{-1}\delta g \right\rangle. \tag{2.40}$$

Now we proceed to calculate the infinitesimal variation of L. On account of (2.36) and (2.40), we obtain

$$\delta L = \sum_i \frac{\partial L}{\partial q^i}\delta q^i + \sum_i \frac{\partial L}{\partial \dot{q}^i}\delta \dot{q}^i + \left\langle g^{-1}\frac{\partial L}{\partial g} - \left(\frac{\partial L}{\partial g}\right)^T g, g^{-1}\delta g \right\rangle + \left\langle \frac{\delta L}{\delta \Pi}, \delta\Pi \right\rangle, \tag{2.41}$$

where we have adopted the left-invariant variation, $g^{-1}\delta g$. Since the variation $\delta\Pi$ is put in the form

$$\delta\Pi = \left[\xi\,,\, g^{-1}\delta g\right] + \frac{d}{dt}\left(g^{-1}\delta g\right) + \sum_{i,j}\left(\frac{\partial\Lambda_j}{\partial q^i} - \frac{\partial\Lambda_i}{\partial q^j}\right)\dot{q}^j\delta q^i + \frac{d}{dt}\Big(\sum_i \Lambda_i\delta q^i\Big), \tag{2.42}$$

by performing the integration by parts and using the formula (2.20), we obtain after calculation the Euler–Lagrange equations given by

$$\frac{d}{dt}\left(\frac{\partial L}{\partial\dot{q}^i}\right) - \frac{\partial L}{\partial q^i}$$
$$= \left\langle \frac{\delta L}{\delta\Pi}, \sum_j K_{ij}\dot{q}^j\right\rangle - \left\langle\frac{\delta L}{\delta\Pi}, [\Pi, \Lambda_i]\right\rangle - \left\langle g^{-1}\frac{\partial L}{\partial g} - \left(\frac{\partial L}{\partial g}\right)^T g, \Lambda_i\right\rangle, \tag{2.43a}$$

$$\frac{d}{dt}\frac{\delta L}{\delta\Pi} = \left[\frac{\delta L}{\delta\Pi}, \Pi\right] - \sum_j\left[\frac{\delta L}{\delta\Pi}, \Lambda_j\right]\dot{q}^j + \left(g^{-1}\frac{\partial L}{\partial g} - \left(\frac{\partial L}{\partial g}\right)^T g\right), \tag{2.43b}$$

where K_{ij} is the curvature tensor defined by

$$K_{ij} := \frac{\partial\Lambda_j}{\partial q^i} - \frac{\partial\Lambda_i}{\partial q^j} - \left[\Lambda_i, \Lambda_j\right], \tag{2.44}$$

and related to κ^c_{ij} given in (1.194), by $K_{ij} = \sum\kappa^c_{ij}R(\boldsymbol{e}_c)$.

Proposition 2.3.1 *The Euler–Lagrange equations for the non-singular configurations are given by* (2.43).

Assume now that L is invariant under the left $SO(3)$ action, i.e., L is rotationally invariant,

$$L(q, \dot{q}, hg, \Pi) = L(q, \dot{q}, g, \Pi) \quad \text{for all } h \in SO(3). \tag{2.45}$$

Note here that Π is left-invariant. Then, this equation with $h = e^{t\eta}$, $\eta \in \mathfrak{so}(3)$, is differentiated with respect to t at $t = 0$ to imply that the right-hand side of (2.37) vanishes for any $\eta \in \mathfrak{so}(3)$, so that one has $\dfrac{\partial L}{\partial g}g^{-1} - g\left(\dfrac{\partial L}{\partial g}\right)^T = 0$, which is also equivalent to

$$g^{-1}\frac{\partial L}{\partial g} - \left(\frac{\partial L}{\partial g}\right)^T g = 0. \tag{2.46}$$

From (2.45), L will reduce to a function $L^*(q, \dot{q}, \Pi)$ on the reduced bundle

$$T(\dot{Q})/SO(3) \cong T(\dot{Q}/SO(3)) \oplus \mathcal{G}, \tag{2.47}$$

where the right-hand side is a Whitney sum bundle, and $\mathcal{G} := \dot{Q} \times_{SO(3)} \mathfrak{so}(3)$ denotes the vector bundle associated with the adjoint action of $SO(3)$ on $\mathfrak{so}(3)$ (see [9, 10] for the reduced bundle). From (2.43) and (2.46), the reduced Euler–Lagrange equations take the form

$$\frac{d}{dt}\left(\frac{\partial L^*}{\partial \dot{q}^i}\right) - \frac{\partial L^*}{\partial q^i} = \left\langle \frac{\delta L^*}{\delta \Pi}, \sum_j K_{ij}\dot{q}^j \right\rangle - \left\langle \frac{\delta L^*}{\delta \Pi}, [\Pi, \Lambda_i] \right\rangle, \tag{2.48a}$$

$$\frac{d}{dt}\frac{\delta L^*}{\delta \Pi} = \left[\frac{\delta L^*}{\delta \Pi}, \Pi\right] - \sum_j \left[\frac{\delta L^*}{\delta \Pi}, \Lambda_j\right]\dot{q}^j. \tag{2.48b}$$

On account of (2.31), Eq. (2.48b) is put also in the form $\dfrac{d}{dt}\dfrac{\delta L^*}{\delta \Pi} = \left[\dfrac{\delta L^*}{\delta \Pi}, \xi\right]$, so that one verifies that

$$\frac{d}{dt}\left(g\frac{\delta L^*}{\delta \Pi}g^{-1}\right) = 0. \tag{2.49}$$

Proposition 2.3.2 *If the Lagrangian is rotationally invariant, the Euler–Lagrange equations for the non-singular configurations reduce to* (2.48), *which are defined on the reduced bundle* (2.47). *Further, the quantity* $g\dfrac{\delta L^*}{\delta \Pi}g^{-1}$ *is conserved.*

We proceed to discuss the equations of motion for the N-body system. We now introduce the vector variable $\boldsymbol{\pi}$ defined through and expressed as

$$\Pi = R(\boldsymbol{\pi}) = \sum \pi^a R(\boldsymbol{e}_a), \quad \boldsymbol{\pi} = \boldsymbol{\Omega} + \sum \boldsymbol{\lambda}_i \dot{q}^i, \tag{2.50}$$

where

$$R(\boldsymbol{\Omega}) = g^{-1}\dot{g}, \quad R(\boldsymbol{\lambda}_i) = \Lambda_i. \tag{2.51}$$

For the spatial many-body system of non-singular configurations, a rotationally-invariant Lagrangian is given, from the kinetic metric (1.196), by

$$L^* = \frac{1}{2}\sum_{i,j} a_{ij}\dot{q}^i\dot{q}^j + \frac{1}{2}\sum_{a,b} A_{ab}\pi^a\pi^b - V(q), \tag{2.52}$$

where a_{ij} and A_{ab} are, respectively, the Riemannian metric and the inertia tensor given by (1.197) and (1.198), and where $V(q)$ denotes a potential function depending on q only. We now rewrite the Lagrangian L^* in terms of Π. On introducing the inertia tensor on $\mathfrak{so}(3)$ by

$$\mathcal{A}_{\sigma(q)} := RA_{\sigma(q)}R^{-1}, \tag{2.53}$$

the Lagrangian is put in the form

$$L^* = \frac{1}{2}\sum_{i,j} a_{ij}\dot{q}^i\dot{q}^j + \frac{1}{2}\langle \Pi, \mathcal{A}_{\sigma(q)}\Pi\rangle - V(q). \tag{2.54}$$

We can apply (2.48) to this Lagrangian to obtain the Euler–Lagrange equations, which do not need to be written down here. What we have to note is that the application of the definition (2.35) to L^* given in (2.54) provides the functional derivative of L^* in the form

$$\frac{\delta L^*}{\delta \Pi} = \mathcal{A}_{\sigma(q)}\Pi = R(A_{\sigma(q)}\boldsymbol{\pi}). \tag{2.55}$$

Further, the quantity $\langle \delta L^*/\delta\Pi, [\Pi, \Lambda_i]\rangle$ is arranged as follows:

$$\begin{aligned}
\langle \mathcal{A}_{\sigma(q)}\Pi, [\Pi, \Lambda_i]\rangle &= \langle [\Lambda_i, \mathcal{A}_{\sigma(q)}\Pi], \Pi\rangle = \langle R(\boldsymbol{\lambda}_i \times A_{\sigma(q)}\boldsymbol{\pi}), R(\boldsymbol{\pi})\rangle \\
&= (\boldsymbol{\lambda}_i \times A_{\sigma(q)}\boldsymbol{\pi}) \cdot \boldsymbol{\pi} = \Lambda_i A_{\sigma(q)}\boldsymbol{\pi} \cdot \boldsymbol{\pi} \\
&= \frac{1}{2}\Lambda_i A_{\sigma(q)}\boldsymbol{\pi} \cdot \boldsymbol{\pi} + \frac{1}{2}\boldsymbol{\pi} \cdot (\Lambda_i A_{\sigma(q)})^T \boldsymbol{\pi} \\
&= \frac{1}{2}\boldsymbol{\pi} \cdot [\Lambda_i, A_{\sigma(q)}]\boldsymbol{\pi},
\end{aligned}$$

where $\boldsymbol{\lambda}_i$ is given in (2.51) and where use has been made of the fact that Λ_i and $A_{\sigma(q)}$ are anti-symmetric and symmetric, respectively.

By the use of this equation, the reduced Euler–Lagrange equations (2.48) are put into the Euler–Lagrange equations of the vector form for the Lagrangian (2.52), and further arranged as

$$\begin{aligned}
&\frac{d}{dt}\dot{q}^i + \sum_{j,k}\Gamma^i_{jk}\dot{q}^j\dot{q}^k + \sum_j a^{ij}\frac{\partial V}{\partial q^j} \\
&= \sum_{j,k} a^{ij}\left(A_{\sigma(q)}\boldsymbol{\pi} \cdot \boldsymbol{\kappa}_{jk}\dot{q}^k + \frac{1}{2}\boldsymbol{\pi} \cdot \left(\frac{\partial A_{\sigma(q)}}{\partial q^j} - \left[\Lambda_j, A_{\sigma(q)}\right]\right)\boldsymbol{\pi}\right),
\end{aligned} \tag{2.56a}$$

$$\frac{d}{dt}\left(A_{\sigma(q)}\boldsymbol{\pi}\right) = A_{\sigma(q)}\boldsymbol{\pi} \times \boldsymbol{\pi} - \sum_i A_{\sigma(q)}\boldsymbol{\pi} \times \boldsymbol{\lambda}_i\dot{q}^i, \tag{2.56b}$$

where $\boldsymbol{\kappa}_{ij}$ are the vectors determined through

$$K_{ij} = R(\boldsymbol{\kappa}_{ij}), \tag{2.57}$$

and where the quantities Γ^i_{jk} denote the Christoffel symbols formed from a_{ij},

$$\Gamma^i_{jk} := \frac{1}{2}\sum_{\ell} a^{i\ell}\left(\frac{\partial a_{\ell k}}{\partial q^j} + \frac{\partial a_{\ell j}}{\partial q^k} - \frac{\partial a_{jk}}{\partial q^\ell}\right), \quad \left(a^{ij}\right) = \left(a_{ij}\right)^{-1}. \tag{2.58}$$

Proposition 2.3.3 *For non-singular configurations, the reduced Euler–Lagrange equations for the Lagrangian* (2.52) *are expressed, in the vector notation, as* (2.56a) *and* (2.56b), *where the right-hand side of Eq.* (2.56a) *describes the sum of a generalized Lorentz force and a generalized centrifugal force. Eq.* (2.56b) *is a (partial) generalization of the Euler equation for a rigid body to one for deformable configurations. These equations of motion are independent of the choice of local sections.*

Proof For the proof of this proposition, we have to explain the generalized Lorentz force and the centrifugal potential and further to show the latter part, (i) the generalization of the Euler equation for a rigid body and (ii) the independence of the choice of local sections. We begin by noting that the conserved quantity $g\dfrac{\delta L^*}{\delta \Pi}g^{-1}$ referred to in Proposition 2.3.2 is indeed the total angular momentum. In fact, a straightforward calculation provides

$$g\frac{\delta L^*}{\delta \Pi}g^{-1} = gR\big(A_{\sigma(q)}(\boldsymbol{\pi})\big)g^{-1} = R\big(gA_{\sigma(q)}(\boldsymbol{\pi})\big) = R(\boldsymbol{L}), \tag{2.59}$$

where the last equality of the above equation is verified, by the use of (1.189), (2.32), (2.50), and (2.51), as follows:

$$\begin{aligned}
\boldsymbol{L} &= \sum_{k=1}^{N-1} \boldsymbol{r}_k \times \dot{\boldsymbol{r}}_k = \sum_k g\boldsymbol{\sigma}_k(q) \times \Big(\dot{g}\boldsymbol{\sigma}_k(q) + g\sum_i \frac{\partial \boldsymbol{\sigma}_k(q)}{\partial q^i}\dot{q}^i\Big) \\
&= g\sum_k \boldsymbol{\sigma}_k(q) \times g^{-1}\dot{g}\boldsymbol{\sigma}_k(q) + gA_{\sigma(q)}A^{-1}_{\sigma(q)}\sum_i\sum_k \boldsymbol{\sigma}_k(q) \times \frac{\partial \boldsymbol{\sigma}_k(q)}{\partial q^i}\dot{q}^i \\
&= gA_{\sigma(q)}\big(\boldsymbol{\Omega} + \sum_i \boldsymbol{\lambda}_i\dot{q}^i\big) = gA_{\sigma(q)}(\boldsymbol{\pi}).
\end{aligned} \tag{2.60}$$

In addition, we can show that (2.56b) is equivalent to the conservation of $\boldsymbol{L} = gA_{\sigma(q)}(\boldsymbol{\pi})$. In this sense, Eq. (2.56b) is a generalization of the Euler equation for a rigid body. In fact, if $\dot{q} = 0$ (or q is constant), then $\boldsymbol{\pi}$ reduces to $\boldsymbol{\Omega}$ with $R(\boldsymbol{\Omega}) = g^{-1}\dot{g}$, so that Eq. (2.56b) reduces to the Euler equation (2.21) for the rigid body.

We turn to the generalized Lorentz force. The quantity $A_{\sigma(q)}\boldsymbol{\pi} \cdot \sum \boldsymbol{\kappa}_{jk}\dot{q}^k$ in the right-hand side of (2.56a) is rewritten as

$$A_{\sigma(q)}\boldsymbol{\pi} \cdot \sum \boldsymbol{\kappa}_{jk}\dot{q}^k = \boldsymbol{L} \cdot g \sum \boldsymbol{\kappa}_{jk}\dot{q}^k = \boldsymbol{L} \cdot \sum \boldsymbol{F}_{jk}\dot{q}^k, \tag{2.61}$$

where Eqs. (2.60) and (1.195) have been used together with the symbol $\boldsymbol{F}_{jk} = \sum F^c_{jk}\boldsymbol{e}_c$. This quantity is a generalization of the Lorentz force $e \sum F_{jk}\dot{q}^k$ for an electron of charge e, where (F_{jk}) is the tensor field for the magnetic flux density (see also (3.1)). Since the curvature F^c_{jk} is vector-valued (or $\mathfrak{so}(3)$-valued), the charge is generalized to be a vector-valued constant of motion. We move ahead to the centrifugal force. The quantity $\frac{1}{2}\boldsymbol{\pi} \cdot A_{\sigma(q)}(\boldsymbol{\pi})$ is rewritten as

$$\frac{1}{2}\boldsymbol{\pi} \cdot A_{\sigma(q)}(\boldsymbol{\pi}) = \frac{1}{2}g\boldsymbol{\pi} \cdot gA_{\sigma(q)}(\boldsymbol{\pi}) = \frac{1}{2}A^{-1}_{g\sigma(q)}\boldsymbol{L} \cdot \boldsymbol{L}. \tag{2.62}$$

This is a generalization of the centrifugal potential, $\ell^2/(2mr^2)$, of a particle with mass m. The quantity $\frac{1}{2}\boldsymbol{\pi} \cdot \big(\frac{\partial A_{\sigma(q)}}{\partial q^j} - [\Lambda_j, A_{\sigma(q)}]\big)\boldsymbol{\pi}$ in the right-hand side of (2.56a) is viewed as the covariant derivative of the centrifugal potential $\frac{1}{2}\boldsymbol{\pi} \cdot A_{\sigma(q)}(\boldsymbol{\pi})$ with respect to $\partial/\partial q^j$, as will be shown below. Hence, the quantity in question stands for a centrifugal force.

We proceed to show that the Euler–Lagrange equations (2.56) are described independently of the choice of local sections by introducing the notion of covariant derivation. Let $\sigma' : U' \to \dot{Q}$ be another local section such that $\sigma'(q) = h(q)\sigma(q)$, $h(q) \in SO(3)$, for $q \in U' \cap U \neq \emptyset$. Further, the $SO(3)$ variable g' is related to g through $g'\sigma'(q) = g\sigma(q)$, so that one has $g' = gh^{-1}$. As in (2.32), the variable ξ' is defined to be $\xi' = g'^{-1}\dot{g}'$ and is shown to transform according to

$$\xi' = h\xi h^{-1} - \dot{h}h^{-1}. \tag{2.63}$$

As was referred to in the previous chapter, the Λ_i and $A_{\sigma(q)}$ transform according to (1.203) and (1.205), respectively. From the transformation law (1.203) for Λ_i and (2.63) for ξ, one can verify that the transformation laws of the curvature tensor and of the variable Π are given by

$$K'_{ij} = h(q)K_{ij}h^{-1}(q), \quad \Pi' = h(q)\Pi h^{-1}(q), \tag{2.64}$$

respectively. Correspondingly, we have the respective transformation laws in the vector form,

$$\boldsymbol{\kappa}'_{ij} = h(q)\boldsymbol{\kappa}_{ij}, \quad \boldsymbol{\pi}' = h(q)\boldsymbol{\pi}. \tag{2.65}$$

Furthermore, we introduce the covariant derivative of $A_{\sigma(q)}$, in view of the quantity appearing in (2.56). From the transformation laws (1.203) and (1.205), it follows that

$$\frac{\partial A'}{\partial q^i} - [\Lambda'_i, A'] = h(q)\left(\frac{\partial A}{\partial q^i} - [\Lambda_i, A]\right)h^{-1}(q), \tag{2.66}$$

where $A = A_{\sigma(q)}$ and $A' = A_{\sigma'(q)}$. This equation means that $\partial A/\partial q^i - [\Lambda_i, A]$ is the covariant derivative of A with respect to $\partial/\partial q^i$. From (2.65) and (2.66), the right-hand side of (2.56a) proves to be independent of the choice of local sections. For Eq. (2.56b), we introduce the covariant derivation along a curve $q(t)$ by

$$\frac{D}{dt} := \frac{d}{dt} - \sum_i \Lambda_i \dot{q}^i, \qquad \Lambda_i = R(\boldsymbol{\lambda}_i). \tag{2.67}$$

Then, Eq. (2.56b) is rewritten as

$$\frac{D}{dt}\left(A_{\sigma(q)}\boldsymbol{\pi}\right) = A_{\sigma(q)}\boldsymbol{\pi} \times \boldsymbol{\pi}. \tag{2.68}$$

As is easily verified, this equation is equivalent to $\frac{D'}{dt}(A'\boldsymbol{\pi}') = A'\boldsymbol{\pi}' \times \boldsymbol{\pi}'$, where D'/dt is defined by replacing Λ_i by Λ'_i in Eq. (2.67). This ends the proof.

We conclude this section with a comment on another expression of the Lagrangian. If we take $(q, g, \dot{q}, \xi)$ as local coordinates of $T(\dot{Q})$, then the Lagrangian (2.52) is put in the form

$$L^* = \frac{1}{2}\sum_{i,j} a_{ij}\dot{q}^i\dot{q}^j + \frac{1}{2}\left(\boldsymbol{\Omega} + \sum_i \boldsymbol{\lambda}_i\dot{q}^i\right)\cdot A_{\sigma(q)}\left(\boldsymbol{\Omega} + \sum_i \boldsymbol{\lambda}_i\dot{q}^i\right) - V(q), \tag{2.69}$$

where Eq. (2.50) has been used. This description of the Lagrangian is usually adopted in the physics literature [53]. Though the equations of motion therein are different in expressions from (2.56), they share the same physical interpretation. According to [53], the term $A_{\sigma(q)}\boldsymbol{\pi} \cdot \boldsymbol{\kappa}_{jk}\dot{q}^k$ or the right-hand side of Eq. (2.61) represents the Coriolis forces and resembles the (generalized) Lorentz force for a charged particle moving in a Yang–Mills field. The quantity in the right-hand side of (2.62) is also referred to as a centrifugal potential. Further, Eq. (2.56b) is viewed as an extended Wong's equation [60, 84] with an extended isospin $A_\sigma(q)\boldsymbol{\pi}$.

While we have so far obtained Lagrange's equations of motion for non-singular configurations, Lagrange's equations of motion can be obtained for singular configurations, which are given in [36], but notations used therein are a little different from those in the present book.

2.4 Hamel's Approach

In this section, we take Hamel's approach [23, 49] to show that the Euler–Lagrange equations so far obtained can also be derived from the Euler–Lagrange equations of the usual form. Let θ^λ and Z_λ be local bases of one-forms and of vector fields on an open subset W of $\mathbb{R}^n$, which are given, respectively, by

$$\theta^\lambda = \sum_\mu A^\lambda_\mu dx^\mu, \quad Z_\lambda = \sum_\mu B^\mu_\lambda \frac{\partial}{\partial x^\mu} \tag{2.70}$$

with

$$\sum_\mu A^\lambda_\mu B^\mu_\nu = \delta^\lambda_\nu, \quad \sum_\lambda A^\lambda_\mu B^\kappa_\lambda = \delta^\kappa_\mu. \tag{2.71}$$

Then one has, after differentiation,

$$d\theta^\lambda = \sum_{\nu<\rho} \gamma^\lambda_{\rho\nu}\theta^\nu \wedge \theta^\rho, \quad \gamma^\lambda_{\rho\nu} := \sum_{\kappa,\mu} \left(\frac{\partial A^\lambda_\mu}{\partial x^\kappa} - \frac{\partial A^\lambda_\kappa}{\partial x^\mu} \right) B^\kappa_\nu B^\mu_\rho, \tag{2.72}$$

where $\gamma^\lambda_{\rho\nu}$'s are called the Hamel symbols [23]. Now, introducing the variable π^λ by

$$\pi^\lambda = \sum_\mu A^\lambda_\mu \dot{x}^\mu, \tag{2.73}$$

and replacing local coordinates $(x, \dot{x})$ in TW by (x, π), we can express the Lagrangian $L(x, \dot{x})$ as

$$\widetilde{L}(x, \pi) = L(x, \dot{x}). \tag{2.74}$$

Proposition 2.4.1 *The Euler–Lagrange equations described in terms of* $(x, \dot{x})$

$$\frac{d}{dt}\frac{\partial L}{\partial \dot{x}^\lambda} - \frac{\partial L}{\partial x^\lambda} = 0 \tag{2.75}$$

are put into the form

$$\frac{d}{dt}\frac{\partial \widetilde{L}}{\partial \pi^\nu} + \sum_{\mu,\rho} \gamma^\mu_{\nu\rho} \frac{\partial \widetilde{L}}{\partial \pi^\mu}\pi^\rho - Z_\nu(\widetilde{L}) = 0, \tag{2.76}$$

which we call Hamel's equations of motion.

Proof A calculation with (2.73) leads to

$$\frac{d}{dt}\Big(\frac{\partial L}{\partial \dot{x}^\lambda}\Big) - \frac{\partial L}{\partial x^\lambda}$$
$$= \sum_{\kappa,\mu} \frac{\partial A^\mu_\lambda}{\partial x^\kappa}\dot{x}^\kappa \frac{\partial \widetilde{L}}{\partial \pi^\mu} + \sum_\mu A^\mu_\lambda \frac{d}{dt}\frac{\partial \widetilde{L}}{\partial \pi^\mu} - \frac{\partial \widetilde{L}}{\partial x^\lambda} - \sum_{\mu,\kappa} \frac{\partial \widetilde{L}}{\partial \pi^\mu}\frac{\partial A^\mu_\kappa}{\partial x^\lambda}\dot{x}^\kappa = 0. \tag{2.77}$$

Multiplying the above equation by B^λ_ν and summing them over λ, we obtain

$$\sum_{\lambda,\kappa,\mu} B^\lambda_\nu \frac{\partial A^\mu_\lambda}{\partial x^\kappa}\dot{x}^\kappa \frac{\partial \widetilde{L}}{\partial \pi^\mu} + \frac{d}{dt}\frac{\partial \widetilde{L}}{\partial \pi^\nu} - \sum_\lambda B^\lambda_\nu \frac{\partial \widetilde{L}}{\partial x^\lambda} - \sum_{\kappa,\mu,\lambda} \frac{\partial \widetilde{L}}{\partial \pi^\mu}\frac{\partial A^\mu_\kappa}{\partial x^\lambda} B^\lambda_\nu \dot{x}^\kappa = 0. \tag{2.78}$$

By using $\dot{x}^\kappa = \sum B^\kappa_\rho \pi^\rho$, we arrange the above equation to obtain

$$\frac{d}{dt}\frac{\partial \widetilde{L}}{\partial \pi^\nu} - \sum_\lambda B^\lambda_\nu \frac{\partial \widetilde{L}}{\partial x^\lambda} + \sum_{\mu,\rho}\sum_{\lambda,\kappa} B^\lambda_\nu B^\kappa_\rho \Big(\frac{\partial A^\mu_\lambda}{\partial x^\kappa} - \frac{\partial A^\mu_\kappa}{\partial x^\lambda}\Big)\frac{\partial \widetilde{L}}{\partial \pi^\mu}\pi^\rho = 0, \tag{2.79}$$

which are rewritten in terms of the Hamel symbols as

$$\frac{d}{dt}\frac{\partial \widetilde{L}}{\partial \pi^\nu} - Z_\nu(\widetilde{L}) + \sum_{\mu,\rho} \gamma^\mu_{\nu\rho}\frac{\partial \widetilde{L}}{\partial \pi^\mu}\pi^\rho = 0. \tag{2.80}$$

This ends the proof.

The Euler–Lagrangian equations of this form have long been known as the equations of motion expressed in terms of the quasi-coordinates [82], but we do not use the indistinct word "quasi-coordinates."

We wish to apply this type of equations to the Lagrangian for a spatial many-body system. As is seen in Sect. 1.8.3, the dq^i, Θ^a and the Y_i, K_a form a local basis of one-forms and of vector fields on $T(\dot{Q})$, respectively. On account of the structure equations (1.96) for the left-invariant one-forms Ψ^a, the exterior derivative of Θ^c proves to be written out as

$$d\Theta^c = -\sum_{a<b} \varepsilon_{abc}\Theta^a\wedge\Theta^b + \sum_i\sum_{a,b}\varepsilon_{abc}\Lambda^b_i\Theta^a\wedge dq^i + \sum_{i<j}\kappa^c_{ij}dq^i\wedge dq^j, \tag{2.81}$$

where κ^c_{ij} are the same as those given in (1.194) and also as the components of the curvature tensor K_{ij} given in (2.44). The exterior derivative of dq^i vanishes. Thus, one finds that the Hamel symbols are put in the form

$$\gamma^c_{ba} = -\varepsilon_{abc}, \quad \gamma^c_{ia} = \sum_b \varepsilon_{abc}\Lambda^b_i, \quad \gamma^c_{ji} = \kappa^c_{ij},$$
$$\gamma^i_{\lambda\mu} = 0 \quad \text{for } \lambda, \mu \in \{a, i\}. \tag{2.82}$$

We now write out the Euler–Lagrange equations (2.80) with the Hamel symbols (2.82) in terms of the local coordinates $q^i, g, \dot{q}^i, \pi^a$ of $T(\dot{Q})$,

$$\frac{d}{dt}\frac{\partial \widetilde{L}}{\partial \dot{q}^i} + \sum_{i,j} \kappa^a_{ji}\frac{\partial \widetilde{L}}{\partial \pi^a}\dot{q}^j + \sum_{a,b}\sum_c \varepsilon_{abc}\Lambda^c_i \frac{\partial \widetilde{L}}{\partial \pi^a}\pi^b - Z_i(\widetilde{L}) = 0, \tag{2.83a}$$

$$\frac{d}{dt}\frac{\partial \widetilde{L}}{\partial \pi^a} - \sum_{b,c} \varepsilon_{abc}\frac{\partial \widetilde{L}}{\partial \pi^b}\pi^c - \sum_{b,i}\sum_c \varepsilon_{acb}\Lambda^c_i \frac{\partial \widetilde{L}}{\partial \pi^b}\dot{q}^i - Z_a(\widetilde{L}) = 0, \tag{2.83b}$$

where $Z_i = Y_i$ and $Z_a = K_a$. By using (2.39) for $Z_i = Y_i$ and $Z_a = K_a$, we obtain

$$Z_i(\widetilde{L}) = \frac{\partial \widetilde{L}}{\partial q^i} - \left\langle g^{-1}\frac{\partial \widetilde{L}}{\partial g} - \left(\frac{\partial \widetilde{L}}{\partial g}\right)^T g, \Lambda_i \right\rangle, \tag{2.84a}$$

$$Z_a(\widetilde{L}) = \left\langle g^{-1}\frac{\partial \widetilde{L}}{\partial g} - \left(\frac{\partial \widetilde{L}}{\partial g}\right)^T g, R(\boldsymbol{e}_a) \right\rangle, \tag{2.84b}$$

respectively. Hence, the Euler–Lagrange equations (2.83) together with the above equations turn out to be equivalent to Eq. (2.43). Moreover, if the Lagrangian is rotationally invariant, one has $g^{-1}\frac{\partial \widetilde{L}}{\partial g} - \left(\frac{\partial \widetilde{L}}{\partial g}\right)^T g = 0$, so that the Euler–Lagrange equations (2.83) reduce to

$$\frac{d}{dt}\frac{\partial \widetilde{L}}{\partial \dot{q}^i} - \frac{\partial \widetilde{L}}{\partial q^i} = \sum_j \frac{\partial \widetilde{L}}{\partial \boldsymbol{\pi}} \cdot \boldsymbol{\kappa}_{ij}\dot{q}^j - \frac{\partial \widetilde{L}}{\partial \boldsymbol{\pi}} \cdot (\boldsymbol{\pi} \times \boldsymbol{\lambda}_i), \tag{2.85a}$$

$$\frac{d}{dt}\frac{\partial \widetilde{L}}{\partial \boldsymbol{\pi}} = \frac{\partial \widetilde{L}}{\partial \boldsymbol{\pi}} \times \boldsymbol{\pi} - \sum_i \frac{\partial \widetilde{L}}{\partial \boldsymbol{\pi}} \times \boldsymbol{\lambda}_i \dot{q}^i, \tag{2.85b}$$

which are also equivalent to Eq. (2.48). See also [50] for the reduced Euler–Lagrange equations on $T(Q)/G$, where we choose $G = SO(3)$ in our case.

So far we have written down the equations of motion by adopting the one-forms dq^i, Θ^a and the vector fields Y_i, K_a. In place of these one-forms and the vector fields, we can use the one-forms $dq^i, \tilde{\omega}^a$ and the vector fields X_i, J_a, which are related by $\tilde{\omega}^a = \sum g_{ab}\Theta^b$ and $J_a = \sum g_{ab}K_b$, whereas we do not pursue this way but refer to [32].

For the spatial three-body system, we have already obtained the expression of $\boldsymbol{\lambda}_i$ and $\boldsymbol{\kappa}_{ij}$ in Sect. 1.9, where $\boldsymbol{\pi}$ is given by

$$\boldsymbol{\pi} = \boldsymbol{\Omega} + \frac{q_q \dot{q}_3 - q_3 \dot{q}_2}{q_1^2 + q_2^2 + q_3^2} \boldsymbol{e}_2, \quad R(\boldsymbol{\Omega}) = g^{-1}\dot{g}.$$

Hence, we can write down the equations of motion for the spatial three-body system, but without going into detail we only refer to [36] for detail. Those equations of motion are very complicated, but they form the foundation for further analysis. For example, we consider small vibrations of a non-linear triatomic molecule near an equilibrium. We can show that if all vibrations have the same period, say 2π, then the non-linear triatomic molecule is allowed to have a perturbed but still periodic motion in the shape space, and further that the periodic motion gives rise to a finite rotation [37]. Though we have already shown that vibrations give rise to rotations on the level of geometry (without reference to equations of motion) in Sect. 1.9, vibrations indeed give rise to rotations on the level of mechanics.

For the planar three-body system, Hamel's equations of motion can be easily written out, but will be given in Appendix 5.8.

2.5 Hamiltonian Mechanics of Many-Body Systems

In this section, we derive the equations of motion for the non-singular configurations on the variational principle in the Hamiltonian formalism [36].

In Sect. 2.3, we have taken $(q, g, \dot{q}, \Pi)$ as local coordinates in $T(\dot{Q})$. For a given Lagrangian $L(q, g, \dot{q}, \Pi)$, we define the generalized momenta $(p, \mathcal{M})$ conjugate to the $(\dot{q}, \Pi)$ to be

$$p_i = \frac{\partial L}{\partial \dot{q}^i}, \quad \mathcal{M} = \frac{\delta L}{\delta \Pi}, \tag{2.86}$$

respectively, and take $(q, g, p, \mathcal{M})$ as local coordinates in the cotangent bundle $T^*(\dot{Q})$. We note here that $\mathcal{M}$ reduces to $R(\boldsymbol{M})$ if the configuration of the system is rigid (see (2.8) and (2.55)). The Hamiltonian $H(q, g, p, \mathcal{M})$ is defined, as usual, to be

$$H(q, g, p, \mathcal{M}) = \sum_i p_i \dot{q}^i + \langle \mathcal{M}, \Pi \rangle - L(q, g, \dot{q}, \Pi). \tag{2.87}$$

We are going to obtain the equations of motion for H on the basis of the variational principle applied to

$$\int_{t_1}^{t_2} \Big(\sum_i p_i \frac{dq^i}{dt} + \langle \mathcal{M}, \Pi \rangle - H(q, g, p, \mathcal{M}) \Big) dt \tag{2.88}$$

with the boundary conditions for infinitesimal variations,

$$\delta q(t_k) = 0, \quad \delta g(t_k) = 0, \quad \delta p(t_k) = 0, \quad \delta \mathcal{M}(t_k) = 0, \quad k = 1, 2. \tag{2.89}$$

We notice here that Hamilton's equations can be obtained without the boundary conditions for δp and $\delta \mathcal{M}$ at $t = t_k$. These boundary conditions are added only for further development of the theory in the Hamiltonian formalism [21]. In a similar manner to that in the Lagrangian formalism, Hamilton's equations of motion are obtained as follows:

$$\dot{q}^i = \frac{\partial H}{\partial p_i}, \quad \Pi = \frac{\delta H}{\delta \mathcal{M}}, \tag{2.90a}$$

$$\dot{p}_i + \frac{\partial H}{\partial q^i} = \left\langle \mathcal{M}, \sum_j K_{ij} \frac{\partial H}{\partial p_j} \right\rangle - \left\langle \mathcal{M}, \left[\frac{\delta H}{\delta \mathcal{M}}, \Lambda_i \right] \right\rangle + \left\langle g^{-1} \frac{\partial H}{\partial g} - \left(\frac{\partial H}{\partial g} \right)^T g, \Lambda_i \right\rangle, \tag{2.90b}$$

$$\dot{\mathcal{M}} = \left[\mathcal{M}, \frac{\delta H}{\delta \mathcal{M}} \right] - \sum_j [\mathcal{M}, \Lambda_j] \frac{\partial H}{\partial p_j} - \left(g^{-1} \frac{\partial H}{\partial g} - \left(\frac{\partial H}{\partial g} \right)^T g \right). \tag{2.90c}$$

Proposition 2.5.1 *The equations of motion for the non-singular configurations in the Hamiltonian formalism are expressed as* (2.90).

Assume that H is invariant under the left $SO(3)$ action,

$$H(q, hg, p, \mathcal{M}) = H(q, g, p, \mathcal{M}) \quad \text{for all } h \in SO(3). \tag{2.91}$$

Note that Π is left-invariant and so is $\mathcal{M}$. Then, for $h = e^{t\eta}$ with $\eta \in \mathfrak{so}(3)$, Eq. (2.91) is differentiated with respect to t at $t = 0$ to bring about the condition of the same form as (2.46),

$$g^{-1} \frac{\partial H}{\partial g} - \left(\frac{\partial H}{\partial g} \right)^T g = 0. \tag{2.92}$$

From the rotational invariance, the H will reduce to a function $H^*(q, p, \mathcal{M})$ on

$$T^*(\dot{Q})/SO(3) \cong T^*(\dot{Q}/SO(3)) \oplus \mathcal{G}^*, \tag{2.93}$$

where $\mathcal{G}^* := \dot{Q} \times_{SO(3)} \mathfrak{so}(3)^*$ denotes the covector bundle associated with the co-adjoint action of $SO(3)$ on $\mathfrak{so}(3)^* \cong \mathfrak{so}(3)$ (see [58, 68] for cotangent bundle reduction). From (2.90) and (2.92), the reduced equations of motion are described as

$$\dot{q}^i = \frac{\partial H^*}{\partial p_i}, \qquad \Pi = \frac{\delta H^*}{\delta \mathcal{M}}, \tag{2.94a}$$

$$\dot{p}_i + \frac{\partial H^*}{\partial q^i} = \left\langle \mathcal{M}, \sum_j K_{ij} \frac{\partial H^*}{\partial p_j} \right\rangle - \left\langle \mathcal{M}, \left[\frac{\delta H^*}{\delta \mathcal{M}}, \Lambda_i \right] \right\rangle, \tag{2.94b}$$

$$\dot{\mathcal{M}} = \left[\mathcal{M}, \frac{\delta H^*}{\delta \mathcal{M}} \right] - \sum_j [\mathcal{M}, \Lambda_j] \frac{\partial H^*}{\partial p_j}. \tag{2.94c}$$

Equation (2.94c) is expressed also as $\dot{\mathcal{M}} = [\mathcal{M}, \xi]$ with $\xi = \Pi - \sum \Lambda_i \dot{q}^i$. From (2.31) and (2.32), one finds that $\xi = g^{-1}\dot{g}$, so that the quantity $g\mathcal{M}g^{-1}$ is shown to be conserved. In fact, one easily verifies that

$$\frac{d}{dt}\left(g\mathcal{M}g^{-1}\right) = g([\xi, \mathcal{M}] + \dot{\mathcal{M}})g^{-1} = 0. \tag{2.95}$$

Hence, $\mathcal{M}$ is put in the form $\mathcal{M} = g^{-1}\zeta g$ with a constant $\zeta \in \mathfrak{so}(3)^* \cong \mathfrak{so}(3)$. This implies that $\mathcal{M}$ is tracking on a coadjoint orbit in each fiber of $\mathcal{G}^*$. According to [58], the symplectic leaves of $T^*(\dot{Q})/G$ are given by $\mathbb{J}^{-1}(\mathcal{O})/G$ for each coadjoint orbit $\mathcal{O}$ in $\mathfrak{g}^*$, where $G = SO(3)$, $\mathfrak{g}^* \cong \mathfrak{so}(3)$, and $\mathbb{J} = g\mathcal{M}g^{-1}$ in the present case. Further, $\mathbb{J}^{-1}(\mathcal{O})/G$ is canonically diffeomorphic to the reduced phase space $\mathbb{J}^{-1}(\mu)/G_\mu$, where $\mu \in \mathcal{O}$ and where G_μ denotes the isotropy subgroup at μ [68].

Proposition 2.5.2 *If the Hamiltonian is rotationally invariant, the equations of motion for the non-singular configurations in the Hamiltonian formalism reduce to* (2.94). *Further, the quantity $g\mathcal{M}g^{-1}$ is conserved.*

For the many-body system, the Hamiltonian to which we apply Eq. (2.94) is obtained from the Lagrangian (2.54) by using (2.87),

$$H^* = \frac{1}{2}\sum_{i,j} a^{ij} p_i p_j + \frac{1}{2}\left\langle \mathcal{M}, \mathcal{A}^{-1}_{\sigma(q)}\mathcal{M} \right\rangle + V(q), \quad \mathcal{A}^{-1}_{\sigma(q)} = RA^{-1}_{\sigma(q)}R^{-1}, \tag{2.96}$$

where p_i and $\mathcal{M}$ defined in (2.86) have been found to be

$$p_i = \sum_j a_{ij}\dot{q}^j, \quad \mathcal{M} = \mathcal{A}_{\sigma(q)}(\Pi), \tag{2.97}$$

and where the functional derivative of H^* is easily calculated as

$$\frac{\delta H^*}{\delta \mathcal{M}} = \mathcal{A}^{-1}_{\sigma(q)}\mathcal{M}.$$

By introducing the vector variable $\boldsymbol{M}$ through

$$\mathcal{M} = R\,(\boldsymbol{M}) \quad \text{with } \boldsymbol{M} := \frac{\partial L^*}{\partial \boldsymbol{\pi}}, \tag{2.98}$$

we obtain

$$\boldsymbol{M} = A_{\sigma(q)}(\boldsymbol{\pi}), \quad \langle \mathcal{M}, \Pi \rangle = \boldsymbol{M} \cdot \boldsymbol{\pi}, \tag{2.99}$$

where we remark that the present variable $\boldsymbol{M}$ is a generalization of the angular momentum in the body frame for the rigid body, as is seen from (2.8). Then, the Hamiltonian (2.96) is put in the form

$$H^* = \frac{1}{2}\sum_{i,j} a^{ij} p_i p_j + \frac{1}{2}\boldsymbol{M} \cdot A^{-1}_{\sigma(q)}\boldsymbol{M} + V(q). \tag{2.100}$$

Correspondingly, Hamilton's equations of motion (2.94) are put into the form

$$\dot{q}^i = \frac{\partial H^*}{\partial p_i}, \quad \pi^a = \frac{\partial H^*}{\partial M_a}, \tag{2.101a}$$

$$\dot{p}_i + \frac{\partial H^*}{\partial q^i} = \boldsymbol{M} \cdot \sum_j \boldsymbol{\kappa}_{ij}\frac{\partial H^*}{\partial p_j} - \boldsymbol{M} \cdot \left(\frac{\partial H^*}{\partial \boldsymbol{M}} \times \boldsymbol{\lambda}_i\right), \tag{2.101b}$$

$$\dot{\boldsymbol{M}} = \boldsymbol{M} \times \frac{\partial H^*}{\partial \boldsymbol{M}} - \sum_i (\boldsymbol{M} \times \boldsymbol{\lambda}_i)\frac{\partial H^*}{\partial p_i}. \tag{2.101c}$$

In correspondence to (2.56), the present Hamilton's equations of motion are written out as

$$\dot{q}^i = \sum_j a^{ij} p_j, \quad \boldsymbol{\pi} = A^{-1}_{\sigma(q)}\boldsymbol{M}, \tag{2.102a}$$

$$\begin{aligned} &\dot{p}_i - \sum_{j,k,\ell} a^{j\ell}\Gamma^k_{\ell i} p_k p_j + \frac{\partial V}{\partial q^i} \\ &= \sum_{j,k} \boldsymbol{M} \cdot \boldsymbol{\kappa}_{ij} a^{jk} p_k - \frac{1}{2}\boldsymbol{M} \cdot \left(\frac{\partial A^{-1}_{\sigma(q)}}{\partial q^i} - \left[\Lambda_i, A^{-1}_{\sigma(q)}\right]\right)\boldsymbol{M}, \end{aligned} \tag{2.102b}$$

$$\dot{\boldsymbol{M}} = \sum_{i,j} a^{ij} p_j \boldsymbol{\lambda}_i \times \boldsymbol{M} + \boldsymbol{M} \times A^{-1}_{\sigma(q)}\boldsymbol{M}, \tag{2.102c}$$

where $\boldsymbol{\kappa}_{ij}$ and $\boldsymbol{\lambda}_i$ were defined in (2.57) and (2.51), respectively, and where $\boldsymbol{M} \cdot \left(\frac{\partial H^*}{\partial \boldsymbol{M}} \times \boldsymbol{\lambda}_i\right)$ has been arranged as $-\frac{1}{2}\boldsymbol{M} \cdot [\Lambda_i, A^{-1}_{\sigma(q)}]\boldsymbol{M}$. The quantity $\dfrac{\partial A^{-1}_{\sigma(q)}}{\partial q^i} - \left[\Lambda_i, A^{-1}_{\sigma(q)}\right]$ is viewed as the covariant derivative of $A^{-1}_{\sigma(q)}$ and transforms in the same manner as in (2.66). In addition, from (2.98) with the transformation law for Π and $\mathcal{A}$, we see that $\mathcal{M}$ and $\boldsymbol{M}$ transform according to

$$\mathcal{M}' = h\mathcal{M}h^{-1}, \quad \boldsymbol{M}' = h\boldsymbol{M},$$

respectively. Hence, the right-hand side of (2.102b) is independent of the choice of local sections. Since $g\mathcal{M}g^{-1} = R(g\boldsymbol{M})$, the total angular momentum $\boldsymbol{L} = g\boldsymbol{M}$ is conserved, which is also a consequence of (2.102c). In particular, if $p = 0$ and if $V = 0$, Eqs. (2.102) reduce to (2.28) for the rigid body. Like (2.56b), Eq. (2.102c) is easily shown to be independent of the choice of local sections. Like (2.61) and (2.62), we can touch on the generalized Lorentz force and the generalized centrifugal potential. In fact, by using $\boldsymbol{L} = g\boldsymbol{M}$, we can verify that

$$\boldsymbol{M} \cdot \sum \boldsymbol{\kappa}_{ij} a^{jk} p_k = g\boldsymbol{M} \cdot g \sum \boldsymbol{\kappa}_{ij} a^{jk} p_k = \boldsymbol{L} \cdot \sum \boldsymbol{F}_{ij} a^{jk} p_k, \tag{2.103}$$

and

$$\boldsymbol{M} \cdot A^{-1}_{\sigma(q)}\boldsymbol{M} = g\boldsymbol{M} \cdot gA^{-1}_{\sigma(q)}\boldsymbol{M} = \boldsymbol{L} \cdot A^{-1}_{g\sigma(q)}(\boldsymbol{L}), \tag{2.104}$$

which show that the generalized Lorentz force and the generalized centrifugal potential are expressed as $\boldsymbol{M} \cdot \sum \boldsymbol{\kappa}_{ij} a^{jk} p_k$ and as $\boldsymbol{M} \cdot A^{-1}_{\sigma(q)}\boldsymbol{M}$, respectively. To compare the covariant derivative in Hamilton's equations of motion with that in the Euler–Lagrange equations, the following equation is of great help:

$$\frac{DA^{-1}}{\partial q^i}\boldsymbol{M} \cdot \boldsymbol{M} + \boldsymbol{\pi} \cdot \frac{DA}{\partial q^i}\boldsymbol{\pi} = 0, \quad \frac{D}{\partial q^i} = \frac{\partial}{\partial q^i} - [\Lambda_i, \cdot], \tag{2.105}$$

which can be proved by the use of the fact that $\Lambda_i A + A\Lambda_i$ is an anti-symmetric matrix. It is easy to show that (2.102b) with the first equation of (2.102a) are equivalent to (2.56a) and that (2.102c) with the second equation of (2.102a) is equivalent to (2.56b).

Thus, in correspondence to Proposition 2.3.3, we have obtained the following:

Proposition 2.5.3 *For non-singular configurations, the reduced Hamilton's equations of motion for the Hamiltonian* (2.100) *are given by* (2.102), *which are described independently of the choice of local sections. Like the reduced Euler–Lagrange equations, the reduced Hamilton's equations of motion contain the generalized Lorentz force and the generalized centrifugal force in the due form.*

In conclusion of this section, we make comments on momentum variables and on the derivation of Hamilton's equations of motion. If we start with the Lagrangian of the form (2.69), we will obtain the momentum variable in the form

$$p_i = \frac{\partial L^*}{\partial \dot{q}^i} = \sum_j a_{ij}\dot{q}^j + \boldsymbol{\lambda}_i \cdot A_{\sigma(q)}\Big(\boldsymbol{\Omega} + \sum_j \boldsymbol{\lambda}_j \dot{q}^j\Big), \tag{2.106}$$

which are usually adopted in the physics literature [53].

We have obtained Hamilton's equations of motion on the variational principle. This method is quite similar to that used in the symplectic formalism. An example of symplectic formalism is given in Appendix 5.9.

Chapter 3
Mechanical Control Systems

This chapter gives typical examples of mechanical control systems, which explain how control inputs are designed to shape mechanical systems.

3.1 Electron Motion in an Electromagnetic Field

Let ϕ and $\boldsymbol{A}$ be the scalar potential and the vector potential of the electric field $\boldsymbol{E}$ and of the magnetic flux density $\boldsymbol{B}$, respectively; $\boldsymbol{E} = -\nabla\phi$, $\boldsymbol{B} = \nabla \times \boldsymbol{A}$. Let q and m denote the charge and the mass of a charged particle, respectively. If its speed is sufficiently smaller than the speed of light, the Lagrangian for the charged particle in the electromagnetic field is given by

$$L = \frac{1}{2}m|\dot{\boldsymbol{r}}|^2 - q\phi + q\boldsymbol{A}\cdot\dot{\boldsymbol{r}},$$

where $\boldsymbol{r} = (x_k)$ and $\dot{\boldsymbol{r}} = (\dot{x}_k)$ are the position vector and the velocity of the charged particle, respectively, in the SI base units [21]. Lagrange's equations of motion are given by and written out as

$$\frac{d}{dt}\frac{\partial L}{\partial \dot{x}_k} - \frac{\partial L}{\partial x_k} = m\frac{d}{dt}\dot{x}_k + q\sum_j\Big(\frac{\partial A_k}{\partial x_j} - \frac{\partial A_j}{\partial x_k}\Big)\dot{x}_j + q\frac{\partial \phi}{\partial x_k} = 0.$$

Since the vector product $\boldsymbol{B} \times \dot{\boldsymbol{r}}$ has the components

$$(\boldsymbol{B} \times \dot{\boldsymbol{r}})_k = \sum_{j,\ell}\varepsilon_{kj\ell}B_j\dot{x}_\ell = \sum_j\Big(\frac{\partial A_k}{\partial x_j} - \frac{\partial A_j}{\partial x_k}\Big)\dot{x}_j, \tag{3.1}$$

T. Iwai, *Geometry, Mechanics, and Control in Action for the Falling Cat*,
Lecture Notes in Mathematics 2289, https://doi.org/10.1007/978-981-16-0688-5_3

the equation of motion for a charged particle in the electromagnetic field is expressed as

$$m\frac{d\boldsymbol{v}}{dt} = q(\boldsymbol{E} + \boldsymbol{v} \times \boldsymbol{B}), \quad \boldsymbol{v} = \dot{\boldsymbol{r}}. \tag{3.2}$$

See Appendix 5.8 for the Kaluza–Klein formulation for the equations of motion.

If the electric field $\boldsymbol{E}$ plays the role of a control input, it takes the form of a vector-valued function of time, $\boldsymbol{E}(t)$. In such a case, the Lagrangian takes the form

$$L = \frac{1}{2}m|\dot{\boldsymbol{r}}|^2 + q\boldsymbol{A} \cdot \dot{\boldsymbol{r}},$$

and Lagrange's equations of motion are put in the form

$$\frac{d}{dt}\frac{\partial L}{\partial \dot{x}_k} - \frac{\partial L}{\partial x_k} = E_k(t), \quad k = 1, 2, 3,$$

which lead to the same equation as (3.2).

For simplicity, the $\boldsymbol{B}$ is assumed to be constant in space and time. Let us introduce an anti-symmetric matrix $\widehat{\boldsymbol{B}}$ by $\widehat{\boldsymbol{B}} = R(\boldsymbol{B})$, where R is the map given in (1.83).

The equation of motion (3.2) is put in the form

$$\frac{d\boldsymbol{v}}{dt} = \frac{q}{m}\boldsymbol{E} - \frac{q}{m}\widehat{\boldsymbol{B}}\boldsymbol{v}. \tag{3.3}$$

As is well known, the solution to this equation is given by

$$\boldsymbol{v}(t) = e^{-\frac{q}{m}t\widehat{\boldsymbol{B}}}\left(\boldsymbol{v}_0 + \int_0^t \frac{q}{m}e^{\frac{q}{m}\tau\widehat{\boldsymbol{B}}}\boldsymbol{E}(\tau)d\tau\right), \quad \boldsymbol{v}_0 = \boldsymbol{v}(0). \tag{3.4}$$

According to the formula (1.88), the matrix exponential $e^{t\widehat{\boldsymbol{b}}}$ with $|\boldsymbol{b}| = 1$ is written out as

$$e^{t\widehat{\boldsymbol{b}}} = \cos t(I - \boldsymbol{b}\boldsymbol{b}^T) + \sin t\,\widehat{\boldsymbol{b}} + \boldsymbol{b}\boldsymbol{b}^T.$$

On setting

$$\boldsymbol{b} = \boldsymbol{B}/B, \quad B = |\boldsymbol{B}|, \quad \omega = \frac{qB}{m},$$

the integrand of the solution (3.4), without the factor $\frac{q}{m}$, is arranged as

$$e^{(q/m)\tau\widehat{\boldsymbol{B}}}\boldsymbol{E}(\tau) = \left(\cos(\omega\tau)(I - \boldsymbol{b}\boldsymbol{b}^T) + \sin(\omega\tau)\widehat{\boldsymbol{b}} + \boldsymbol{b}\boldsymbol{b}^T\right)\boldsymbol{E}(\tau). \tag{3.5}$$

We are interested in whether one can halt the motion at a certain time t_0, $\boldsymbol{v}(t_0) = 0$, or not, by adjusting $\boldsymbol{E}(t)$. We assume here that the control input is subject to the constraint $\boldsymbol{E}(t) \cdot \boldsymbol{B} = 0$. Then, on account of (3.5), Eq. (3.4) reduces to

$$\boldsymbol{v}(t) = e^{-\frac{q}{m}t\hat{\boldsymbol{B}}}\Big(\boldsymbol{v}_0 + \frac{q}{m}\int_0^t \big(\cos(\omega\tau)\,\boldsymbol{E}(\tau) + \sin(\omega\tau)\,(\boldsymbol{B}\times\boldsymbol{E}(\tau))/B\big)d\tau\Big), \tag{3.6}$$

where use has been made of $\boldsymbol{b}^T\boldsymbol{E} = (\boldsymbol{B}/B)\cdot\boldsymbol{E} = 0$ on account of the assumption. We note that the constraint $\boldsymbol{E}(\tau)\cdot\boldsymbol{B} = 0,\ 0 \le \tau \le t$, means that the vectors $\boldsymbol{E}(\tau)$ and $\boldsymbol{B}\times\boldsymbol{E}(\tau)$ are orthogonal to the fixed vector $\boldsymbol{B}$. This implies that any control works only perpendicularly to the fixed axis $\boldsymbol{B}$. Therefore, if the initial velocity $\boldsymbol{v}_0$ has a non-vanishing component parallel to $\boldsymbol{B}$, i.e., if $\boldsymbol{v}_0\cdot\boldsymbol{B} \neq 0$, then one has

$$\boldsymbol{v}_0 + \frac{q}{m}\int_0^t \big(\cos(\omega\tau)\,\boldsymbol{E}(\tau) + \sin(\omega\tau)\,(\boldsymbol{B}\times\boldsymbol{E}(\tau))/B\big)d\tau \neq 0 \quad \text{for} \quad \forall \boldsymbol{E}(\tau),$$

so that $\boldsymbol{v}(t) \neq 0$ for any $t \in \mathbb{R}$. In other words, the charged particle cannot halt by any control input $\boldsymbol{E}(t)$, if $\boldsymbol{E}(t)\cdot\boldsymbol{B} = 0$.

We give here a brief review of controllability of control systems. For simplicity, we consider the linear inhomogeneous differential equation

$$\frac{d\boldsymbol{x}}{dt} = A(t)\boldsymbol{x} + \boldsymbol{b}(t), \qquad \boldsymbol{x} \in \mathbb{R}^n,$$

where $A(t) \in \mathbb{R}^{n\times n}$ and where $\boldsymbol{b}(t) \in \mathbb{R}^n$ is considered as a control input. A question in the control theory is whether one can design a control input $\boldsymbol{b}(t)$ so that the initial state $\boldsymbol{x}(0) \neq 0$ may be transferred to the origin at a certain time $\tau > 0$, i.e., $\boldsymbol{x}(\tau) = 0$. If such a control input exists, the system is called controllable. If not, the system is called uncontrollable.

We further assume that $A(t)$ is constant in t and $\boldsymbol{b}(t)$ takes the form $\boldsymbol{b}(t) = B\boldsymbol{u}(t)$ on account of the degree of freedom for the control input, where $B \in \mathbb{R}^{n\times m}$, $\boldsymbol{u} \in \mathbb{R}^m$ with $m \le n$. The following theorem is well known (for proof, see [69], for example).

Proposition 3.1.1 *The linear control system on $\mathbb{R}^n$,*

$$\frac{d\boldsymbol{x}}{dt} = A\boldsymbol{x} + B\boldsymbol{u}(t), \qquad A \in \mathbb{R}^{n\times n},\ B \in \mathbb{R}^{n\times m},\ \boldsymbol{u} \in \mathbb{R}^m, \tag{3.7}$$

is controllable, if and only if

$$\operatorname{rank}(B, AB, A^2B, \cdots, A^{n-1}B) = n. \tag{3.8}$$

In order to understand the meaning of controllability from the viewpoint of differential geometry, we introduce the vector fields

$$X = \sum_{i,j} a_{ij} x^j \frac{\partial}{\partial x^i}, \quad Y_r = \sum_k b_{kr} \frac{\partial}{\partial x^k}, \quad i, j, k = 1, \ldots, n, \; r = 1, \ldots, m,$$

where we call X and Y_r a drift vector field and control vector fields, respectively. The control equation (3.7) is put in the form

$$\frac{dx}{dt} = X + \sum u_r Y_r, \quad x \in \mathbb{R}^n.$$

Solutions to this equation take the same form as (3.4), which implies that during the time-evolution, $\exp(tX)$, of the drift vector field X, the control input $\sum u_r Y_r$ interferes with the drift. Then, the interference between the drift vector field X and the control input Y_r may generate new vector fields for control. In fact, a straightforward calculation provides

$$[X, Y_r] = -\sum (AB)_{ir} \frac{\partial}{\partial x^i},$$

$$[X, [X, Y_r]] = \sum (A^2 B)_{ir} \frac{\partial}{\partial x^i}, \quad etc.$$

We observe that the coefficient matrices in the right-hand sides of the above equations are the same as submatrices of the matrix given in (3.8). It then turns out that the system is controllable if and only if the derived flag of distributions becomes of full rank. The present viewpoint has a generic meaning: Interference among degrees of freedom may give rise to a new degree of freedom. With this generic viewpoint in mind, revisiting Eq. (1.176), we see that the equation $[X_i, X_j] = -\sum F_{ij}^c J_c$ means that interference among vibrational degrees of freedom gives rise to rotational degrees of freedom. In contrast to this, interference among rotational vectors is restricted to themselves, which is expressed by $[J_a, J_b] = -\sum \varepsilon_{abc} J_c$. From a mathematical point of view, the controllability is a consequence of Chow's theorem [78]. See [11, 63] for Chow's theorem.

We now apply Proposition 3.1.1 to our system (3.3). We first assume that $\boldsymbol{B} \cdot \boldsymbol{E}(t) = 0$. On taking the Cartesian coordinates (x_1, x_2, x_3) in a manner such that $\boldsymbol{B} = B_3 \boldsymbol{e}_3$, the control input is put in the form

$$\boldsymbol{E} = u_1 \boldsymbol{e}_1 + u_2 \boldsymbol{e}_2 = (\boldsymbol{e}_1, \boldsymbol{e}_2) \begin{pmatrix} u_1 \\ u_2 \end{pmatrix}.$$

In the present coordinate system, we have

$$A = -\frac{q}{m} \widehat{\boldsymbol{B}} = \frac{q}{m} \begin{pmatrix} 0 & B_3 & 0 \\ -B_3 & 0 & 0 \\ 0 & 0 & 0 \end{pmatrix}, \quad B = \begin{pmatrix} 1 & 0 \\ 0 & 1 \\ 0 & 0 \end{pmatrix}.$$

A straightforward calculation gives

$$AB = \frac{q}{m}\begin{pmatrix} 0 & B_3 \\ -B_3 & 0 \\ 0 & 0 \end{pmatrix}, \quad A^2B = \frac{q}{m}\begin{pmatrix} -B_3^2 & 0 \\ -0 & -B_3^2 \\ 0 & 0 \end{pmatrix}$$

This implies that the controllability matrix (B, AB, A^2B) is of rank two, so that the system is uncontrollable.

In contrast to this, if we choose the electric field of the form

$$\boldsymbol{E} = u_1\boldsymbol{e}_1 + u_3\boldsymbol{e}_3 = (\boldsymbol{e}_1, \boldsymbol{e}_3)\begin{pmatrix} u_1 \\ u_3 \end{pmatrix},$$

then, for

$$A = -\frac{q}{m}\widehat{\boldsymbol{B}}, \quad B = \begin{pmatrix} 1 & 0 \\ 0 & 0 \\ 0 & 1 \end{pmatrix},$$

we obtain

$$AB = \frac{q}{m}\begin{pmatrix} 0 & 0 \\ -B_3 & 0 \\ 0 & 0 \end{pmatrix}, \quad A^2B = \frac{q}{m}\begin{pmatrix} -B_3^2 & 0 \\ 0 & 0 \\ 0 & 0 \end{pmatrix}.$$

This implies that the controllability matrix (B, AB, A^2B) is of full rank, so that the system is controllable, though the number of the control vectors is less than the dimension of the space.

For confirmation of the controllability, we now give an explicit form of a control input $\boldsymbol{u}(t)$ which makes the velocity $\boldsymbol{v}(t)$ vanish at a certain time t_0. Setting $\boldsymbol{E}(\tau) = E\boldsymbol{u}(\tau)$, where $E \in \mathbb{R}^{3\times m}$, $1 \le m \le 3$, $\boldsymbol{u}(\tau) \in \mathbb{R}^m$ and further, by setting

$$\boldsymbol{u}(\tau) = E^T e^{\frac{q}{m}\tau\widehat{\boldsymbol{B}}^T}\boldsymbol{E}_0, \quad F(t) = \int_0^t \frac{q}{m} e^{\frac{q}{m}\tau\widehat{\boldsymbol{B}}} E E^T e^{\frac{q}{m}\tau\widehat{\boldsymbol{B}}^T} d\tau,$$

where $\boldsymbol{E}_0 = \boldsymbol{E}(0)$ is a constant vector to be determined soon, we rewrite the solution (3.4) as

$$\boldsymbol{v}(t) = e^{-\frac{q}{m}t\widehat{\boldsymbol{B}}}\big(\boldsymbol{v}_0 + F(t)\boldsymbol{E}_0\big).$$

If $F(t)$ is invertible, the equation $\boldsymbol{v}_0 + F(t)\boldsymbol{E}_0 = 0$ has a solution $\boldsymbol{E}_0$ for any $\boldsymbol{v}_0$, so that the system proves to be controllable. Since $F(t)$ is a symmetric matrix, it is invertible if and only if it is positive-definite. Our last task is to show that $F(t)$ is positive-definite. For any $\boldsymbol{\xi} \in \mathbb{R}^3$, one has

$$\langle \boldsymbol{\xi}, F(t)\boldsymbol{\xi} \rangle = \frac{q}{m} \int_0^t \left\langle E^T e^{\frac{q}{m}\tau \widehat{\boldsymbol{B}}^T} \boldsymbol{\xi}, E^T e^{\frac{q}{m}\tau \widehat{\boldsymbol{B}}^T} \boldsymbol{\xi} \right\rangle d\tau \geq 0.$$

The equality holds if and only if $E^T e^{\frac{q}{m}\tau \widehat{\boldsymbol{B}}^T} \boldsymbol{\xi} = 0$ for any τ with $0 \leq \tau \leq t$. We consider the transposed equation $\boldsymbol{\xi}^T e^{\frac{q}{m}\tau \widehat{\boldsymbol{B}}} E = 0$, which is written out as

$$\boldsymbol{\xi}^T \left(I + \frac{q\tau}{m} \widehat{\boldsymbol{B}} + \frac{1}{2}\left(\frac{q\tau}{m}\right)^2 \widehat{\boldsymbol{B}}^2 + \frac{1}{3!}\left(\frac{q\tau}{m}\right)^3 \widehat{\boldsymbol{B}}^3 + \cdots \right) E = 0.$$

On account of the controllability condition (3.8) with $A = \widehat{\boldsymbol{B}}$ and $B = E$, the above equation implies that $\boldsymbol{\xi} = 0$, so that $F(t)$ is positive-definite.

3.2 The Inverted Pendulum on a Cart

We consider the inverted pendulum on a cart, which consists of a massless rod and a mass m at the top end of the rod, where the rod is joined to the cart with a frictionless joint, and where the length of the rod is ℓ and the mass of the cart is M. A control force u is applied to the cart in the horizontal direction (Fig. 3.1). Let θ denote the angle of the rod with the vertical line, and s the position of the cart. Let (x, y) denote the Cartesian coordinates of the position of the mass m. Then, x and y are described as

$$x = s + \ell \sin\theta, \quad y = \ell \cos\theta.$$

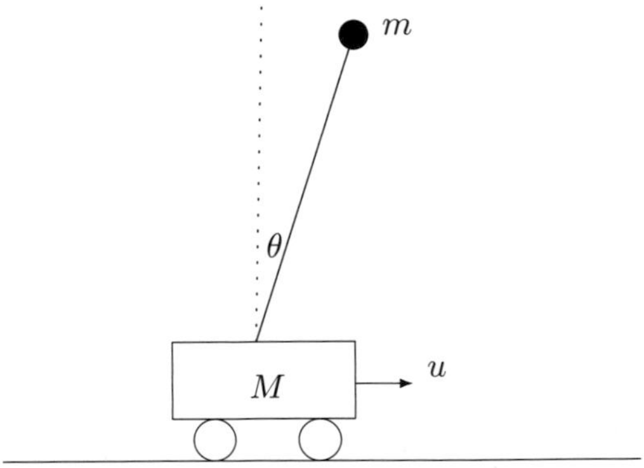

Fig. 3.1 A schematic description of the inverted pendulum on a cart

The total kinetic energy and the potential energy of this system are given by

$$T = \frac{1}{2}m(\dot{x}^2 + \dot{y}^2) + \frac{1}{2}M\dot{s}^2, \quad V = mg\ell\cos\theta,$$

respectively, where g is the gravitational acceleration. In terms of the coordinates (s, θ) and the velocities $(\dot{s}, \dot{\theta})$, the Lagrangian is expressed as

$$L = T - V = \frac{1}{2}\Big((m + M)\dot{s}^2 + 2m\ell\cos\theta\,\dot{s}\dot{\theta} + m\ell^2\,\dot{\theta}^2\Big) - mg\ell\cos\theta.$$

Lagrange's equations of motion with the control input u are

$$\frac{d}{dt}\left(\frac{\partial L}{\partial\dot{\theta}}\right) - \frac{\partial L}{\partial\theta} = m\ell\cos\theta\,\ddot{s} + m\ell^2\ddot{\theta} - mg\ell\sin\theta = 0,$$
$$\frac{d}{dt}\left(\frac{\partial L}{\partial\dot{s}}\right) - \frac{\partial L}{\partial s} = (m + M)\ddot{s} - m\ell\sin\theta\,\dot{\theta}^2 + m\ell\cos\theta\,\ddot{\theta} = u.$$

If $u = 0$, that is, if no force is applied to the cart, the state $(s, \theta, \dot{s}, \dot{\theta}) = (a, 0, 0, 0)$ is an equilibrium, where a is an arbitrary constant. As is well known, this state is unstable. Put another way, the rod is likely to swing downward. We show that there is a way to stabilize this state (or to keep the rod upheld) by applying a suitable control input u. We employ a linearization method by setting $a = 0$ without loss of generality. Let us assume that $(s, \theta, \dot{s}, \dot{\theta})$ be small enough, i.e., $(s, \theta, \dot{s}, \dot{\theta})$ are variables taking values in a vicinity of $(0, 0, 0, 0)$. Then, the equations of motion are linearized as

$$\begin{aligned}(m + M)\ddot{s} + m\ell\ddot{\theta} &= u,\\ m\ell\,\ddot{s} + m\ell^2\ddot{\theta} - mg\ell\theta &= 0,\end{aligned} \tag{3.9}$$

which are arranged as

$$\ddot{s} = -\frac{mg\,\theta}{M} + \frac{u}{M},$$
$$\ddot{\theta} = \frac{g(M + m)\,\theta}{\ell M} - \frac{u}{\ell M}.$$

Introducing the velocities, v_s and v_θ, associated with the variables s and θ, respectively, we put the linearized equations of motion in the form

$$\frac{d\boldsymbol{x}}{dt} = A\boldsymbol{x} + \boldsymbol{b}u(t), \tag{3.10}$$

where

$$x = \begin{pmatrix} s \\ \theta \\ v_s \\ v_\theta \end{pmatrix}, \quad A = \begin{pmatrix} & & 1 & \\ & & & 1 \\ 0 & a_3 & & \\ 0 & a_4 & & \end{pmatrix}, \quad b = \begin{pmatrix} 0 \\ 0 \\ b_3 \\ b_4 \end{pmatrix}, \tag{3.11}$$

$$a_3 = -\frac{mg}{M}, \quad a_4 = \frac{g(M+m)}{\ell M}, \quad b_3 = \frac{1}{M}, \quad b_4 = -\frac{1}{\ell M}. \tag{3.12}$$

We wish to design a feedback control of the form $u(t) = \boldsymbol{k} \cdot \boldsymbol{x}(t)$, where $\boldsymbol{k} = (k_j) \in \mathbb{R}^4$ is a constant vector. Then, Eq. (3.10) turns into

$$\frac{d\boldsymbol{x}}{dt} = (A + \boldsymbol{b}\boldsymbol{k}^T)\boldsymbol{x}, \tag{3.13}$$

where the new coefficient matrix is written out as

$$A + \boldsymbol{b}\boldsymbol{k}^T = \begin{pmatrix} 0 & 0 & 1 & 0 \\ 0 & 0 & 0 & 1 \\ k_1 b_3 & a_3 + k_2 b_3 & k_3 b_3 & k_4 b_3 \\ k_1 b_4 & a_4 + k_2 b_4 & k_3 b_4 & k_4 b_4 \end{pmatrix}. \tag{3.14}$$

If we can choose the vector $\boldsymbol{k}$ so that all of the eigenvalues of the new matrix $A+\boldsymbol{b}\boldsymbol{k}^T$ may have negative real parts, any solution to (3.13) tends to the origin, $\boldsymbol{x}(t) \to 0$, as $t \to \infty$. Thus, the origin is stabilized. Now we calculate the characteristic polynomial for $A + \boldsymbol{b}\boldsymbol{k}^T$ to obtain

$$\begin{aligned} f(\lambda) :&= |\lambda I_4 - (A + \boldsymbol{b}\boldsymbol{k}^T)| \\ &= \lambda^4 - (k_3 b_3 + k_4 b_4)\lambda^3 - (a_4 + k_2 b_4 + k_1 b_3)\lambda^2 \\ &\quad + (a_4 b_3 k_3 - a_3 b_4 k_3)\lambda + a_4 b_3 k_1 - a_3 b_4 k_1, \end{aligned} \tag{3.15}$$

where I_4 denotes the 4×4 identity. We show that k_j can be adjusted so that all the eigenvalues of $A + \boldsymbol{b}\boldsymbol{k}^T$ may have negative real parts. Let λ_j, $j = 1, \ldots, 4$, be a set of negative real numbers and/or conjugate complex numbers with negative real parts. Then, the polynomial having these λ_j as zeros is expressed and expanded as

$$(\lambda - \lambda_1) \cdots (\lambda - \lambda_4) = \lambda^4 + c_3\lambda^3 + c_2\lambda^2 + c_1\lambda + c_0. \tag{3.16}$$

We equate c_j with the corresponding coefficients of (3.15) to obtain

$$c_3 = -(k_3 b_3 + k_4 b_4), \quad c_2 = -(a_4 + k_2 b_4 + k_1 b_3),$$

$$c_1 = a_4 b_3 k_3 - a_3 b_4 k_3, \quad c_0 = a_4 b_3 k_1 - a_3 b_4 k_1.$$

For c_j as given parameters, these equations determine k_j to be

$$k_1 = \frac{\ell M}{g} c_0, \quad k_2 = \ell M c_2 + g(M+m) + \ell k_1,$$

$$k_3 = \frac{\ell M}{g} c_1, \quad k_4 = \ell M c_3 + k_3 \ell,$$

and thereby the inverted pendulum is linearly stabilized.

The procedure we have taken for stabilizing an unstable equilibrium is known as the method of pole assignment (or placement) in the theory of linear control systems (for proof, see [69], for example).

Proposition 3.2.1 *For a linear control system on* $\mathbb{R}^n$,

$$\frac{d\boldsymbol{x}}{dt} = A\boldsymbol{x} + B\boldsymbol{u}(t), \quad A \in \mathbb{R}^{n\times n}, \ B \in \mathbb{R}^{n\times m}, \ \boldsymbol{u} \in \mathbb{R}^m, \tag{3.17}$$

there exists a feedback control $\boldsymbol{u} = K\boldsymbol{x}$ *with* $K \in \mathbb{R}^{m\times n}$ *such that the matrix* $A+BK$ *has arbitrarily assigned eigenvalues, if and only if the pair* (A, B) *is controllable, i.e.,*

$$\text{rank}(B, AB, A^2B, \cdots, A^{n-1}B) = n. \tag{3.18}$$

For the control system (3.10), the controllability matrix is evaluated as

$$(\boldsymbol{b}, A\boldsymbol{b}, A^2\boldsymbol{b}, A^3\boldsymbol{b}) = \begin{pmatrix} 0 & b_3 & 0 & a_3b_4 \\ 0 & b_4 & 0 & a_1b_4 \\ b_3 & 0 & a_3b_4 & 0 \\ b_4 & 0 & a_1b_4 & 0 \end{pmatrix},$$

which is of full rank indeed, as is wanted.

A necessary and sufficient condition is already known for all zeros of a polynomial to have a negative real part. Let $f(x) = p_0 + p_1x + \cdots + p_nx^n$ with $p_0, \cdots, p_n \in \mathbb{R}$. If $p_n > 0$, then all the roots of $f(x)$ have a negative real part if and only if the elements of the Routh array are all positive (for the definition of the Routh array, see [69]). In particular, for the polynomial $\lambda^4 + c_3\lambda^3 + c_2\lambda^2 + c_1\lambda + c_0$ given in (3.16), it is known [69] that all the roots of this polynomial have a negative real part if and only if

$$c_0 > 0, \ c_1 > 0, \ c_2 > 0, \ c_3 > 0, \ c_1c_2c_3 - c_0c_3^2 - c_1^2 > 0. \tag{3.19}$$

In order to gain a mechanical understanding of the pole assignment procedure which we have applied above, we rewrite Eq. (3.9) as

$$M\frac{d^2\boldsymbol{r}}{dt^2} = G\boldsymbol{r} + u\boldsymbol{e}_1, \tag{3.20}$$

where $\boldsymbol{r}$, M, and G are the position vector and the matrices defined to be

$$\boldsymbol{r} = \begin{pmatrix} s \\ \theta \end{pmatrix}, \quad M = \begin{pmatrix} m + M & m\ell \\ m\ell & m\ell^2 \end{pmatrix}, \quad G = \begin{pmatrix} 0 & 0 \\ 0 & mg\ell \end{pmatrix}, \tag{3.21}$$

respectively. On introducing the velocity vector by $\boldsymbol{v} = \dot{\boldsymbol{r}}$ and putting the already-chosen control input in the form

$$u(t) = \begin{pmatrix} \boldsymbol{k}_{\mathrm{u}} \\ \boldsymbol{k}_{\mathrm{d}} \end{pmatrix} \cdot \begin{pmatrix} \boldsymbol{r} \\ \boldsymbol{v} \end{pmatrix}, \quad \boldsymbol{k}_{\mathrm{u}}, \boldsymbol{k}_{\mathrm{d}} \in \mathbb{R}^2,$$

Equation (3.13) takes the form

$$\frac{d}{dt}\begin{pmatrix} \boldsymbol{r} \\ \boldsymbol{v} \end{pmatrix} = \begin{pmatrix} 0 & I_2 \\ M^{-1}G + M^{-1}\boldsymbol{e}_1\boldsymbol{k}_{\mathrm{u}}^T & M^{-1}\boldsymbol{e}_1\boldsymbol{k}_{\mathrm{d}}^T \end{pmatrix} \begin{pmatrix} \boldsymbol{r} \\ \boldsymbol{v} \end{pmatrix}, \tag{3.22}$$

where I_2 denotes the 2×2 identity. Equivalently, the present equation is put into the form

$$\frac{d^2\boldsymbol{r}}{dt^2} = M^{-1}\boldsymbol{e}_1\boldsymbol{k}_{\mathrm{d}}^T\frac{d\boldsymbol{r}}{dt} + M^{-1}(G + \boldsymbol{e}_1\boldsymbol{k}_{\mathrm{u}}^T)\boldsymbol{r}. \tag{3.23}$$

This shows that the designed control input has modified the uncontrolled system $\frac{d^2\boldsymbol{r}}{dt^2} = M^{-1}G\boldsymbol{r}$ into a dynamical system which has the damping term and the moderated repulsive force. In fact, the coefficient matrix of the first term of the right-hand side of (3.23) has the property

$$\det(M^{-1}\boldsymbol{e}_1\boldsymbol{k}_{\mathrm{d}}^T) = 0, \quad \operatorname{tr}\left(M^{-1}\boldsymbol{e}_1\boldsymbol{k}_{\mathrm{d}}^T\right) = -c_3 < 0,$$

which shows that the matrix in question has zero and a negative eigenvalue, since the coefficients c_j of (3.16) are all positive (see (3.19)). As for the coefficient matrix of the second term in the right-hand side of (3.23), we obtain

$$\det(M^{-1}(G + \boldsymbol{e}_1\boldsymbol{k}_{\mathrm{u}}^T)) = c_0 > 0,$$
$$\operatorname{tr}\left(M^{-1}(G + \boldsymbol{e}_1\boldsymbol{k}_{\mathrm{u}}^T)\right) = -c_2 - \frac{g(M + m + m\ell)}{\ell M} < 0,$$

which shows that the real eigenvalues of the present matrix are both negative or that the real part of each eigenvalue is negative. This is a remarkable modification of the corresponding property for the uncontrolled system,

$$\det(M^{-1}G) = 0, \quad \operatorname{tr}(M^{-1}G) = a_4 > 0.$$

To reexamine the controlled system from the viewpoint of energy balance, we rewrite (3.23) as

$$M\frac{d^2\boldsymbol{r}}{dt^2} - G\boldsymbol{r} = \boldsymbol{e}_1\boldsymbol{k}_{\mathrm{d}}^T\frac{d\boldsymbol{r}}{dt} + \boldsymbol{e}_1\boldsymbol{k}_{\mathrm{u}}^T\boldsymbol{r}. \tag{3.24}$$

Taking the inner product of both sides with the velocity vector $\frac{d\boldsymbol{r}}{dt}$, we obtain

$$\frac{d}{dt}\Big(\frac{1}{2}\frac{d\boldsymbol{r}}{dt}\cdot M\frac{d\boldsymbol{r}}{dt} - \frac{1}{2}\boldsymbol{r}\cdot G\boldsymbol{r}\Big) = \Big(\boldsymbol{e}_1\cdot\frac{d\boldsymbol{r}}{dt}\Big)\big(\boldsymbol{k}_{\mathrm{d}}^T\cdot\dot{\boldsymbol{r}} + \boldsymbol{k}_{\mathrm{u}}\cdot\boldsymbol{r}\big) = (\boldsymbol{e}_1\cdot\dot{\boldsymbol{r}})u. \tag{3.25}$$

Since the kinetic energy and the potential energy of the uncontrolled system are

$$T = \frac{1}{2}\frac{d\boldsymbol{r}}{dt}\cdot M\frac{d\boldsymbol{r}}{dt}, \quad V = -\frac{1}{2}\boldsymbol{r}\cdot G\boldsymbol{r},$$

the above equation means that the increase in the total energy of the system is equal to the supplied work.

So far the control for the stabilization of an unstable equilibrium point has been designed for the linearized mechanical system by shaping the equations of motion. The idea of shaping the kinetic and/or the potential energy can be applied to non-linear mechanical systems, and has been established as the method of controlled Lagrangians [6], which has been applied to the stabilization for the (non-linear) inverted pendulum [4, 5].

In comparison to the control theoretic method for stabilization of an unstable equilibrium, the unstable equilibrium of the inverted pendulum can be stabilized, if the point of suspension oscillates in the vertical direction [43]. Stabilization can be discussed within non-autonomous linear dynamical systems [2]. In a similar manner, the stable equilibrium of the pendulum can be linearly instable, if described as a linear non-autonomous system with the length of the rod changing in periodic and piecewise-constant manner [2].

3.3 Port-Hamiltonian Systems

To get an idea of port-Hamiltonian systems, we start with an LC circuit as a simple example, where the LC circuit is an electric circuit consisting of an inductor and a capacitor connected together, where the inductance and the capacitance are usually denoted by L and C, respectively. As is well known, according to Kirchhoff's current law, the circuit is subject to the differential equation

$$L\frac{d^2q}{dt^2} + \frac{1}{C}q = 0, \tag{3.26}$$

where q is the charge stored in the capacitor. This equation is viewed as Lagrange's equation associated with the Lagrangian

$$\mathcal{L} = \frac{1}{2}L\dot{q}^2 - \frac{1}{2C}q^2.$$

On introducing the momentum variable $p = \partial\mathcal{L}/\partial\dot{q}$, the corresponding Hamiltonian H is expressed as

$$H = \frac{1}{2L}p^2 + \frac{1}{2C}q^2.$$

The circuit equation is now expressed as Hamilton's equations of motion on the phase space,

$$\frac{dq}{dt} = \frac{\partial H}{\partial p}, \quad \frac{dp}{dt} = -\frac{\partial H}{\partial q}.$$

Since H is a constant of motion, solutions to the above equations form a family of ellipses around the origin, and the origin is a stable equilibrium.

If a voltage v is impressed into this circuit, the circuit equation is changed into

$$L\frac{d^2q}{dt^2} + \frac{1}{C}q = v. \tag{3.27}$$

The corresponding Hamilton's equations of motion then take the form

$$\frac{dq}{dt} = \frac{\partial H}{\partial p}, \quad \frac{dp}{dt} = -\frac{\partial H}{\partial q} + v. \tag{3.28}$$

If v is constant in t, $v(t) = v_0$, then Eq. (3.27) has a particular solution $q(t) = Cv_0$. This implies that the equilibrium point is shifted from $(0, 0)$ to $(Cv_0, 0)$ in the phase space. In other words, Eq. (3.28) is rewritten as Hamilton's equations of motion associated with the modified Hamiltonian

$$H' = \frac{1}{2L}p^2 + \frac{1}{2C}(q - Cv_0)^2. \tag{3.29}$$

If we take v_0 as a control input, the control is viewed as serving to shape a new potential function.

We are now in a position to introduce port-Hamiltonian systems, which are open dynamical systems, like (3.28), interacting with their environment through ports [73]. In place of v, we take into account a control input denoted by u, and further introduce the Hamiltonian vector field and a control vector field by

$$X_H = \frac{\partial H}{\partial p}\frac{\partial}{\partial q} - \frac{\partial H}{\partial q}\frac{\partial}{\partial p}, \quad Y = \frac{\partial}{\partial p},$$

respectively. Then, the vector field defined by the right-hand side of (3.28) is described as

$$X_H + uY. \tag{3.30}$$

The energy balance equation is expressed and arranged as

$$\frac{dH}{dt} = \dot{q}\frac{\partial H}{\partial q} + \dot{p}\frac{\partial H}{\partial p} = u\frac{p}{L}, \tag{3.31}$$

which implies that the system output is defined to be $\frac{p}{L} = \frac{\partial H}{\partial p} = Y(H)$. In view of this, the system of equations

$$\frac{d\tilde{p}}{dt} = X_H + uY, \quad y = Y(H) \tag{3.32}$$

is called a port-Hamiltonian system with the system output y, where $\tilde{p}$ in the left-hand side stands for a point (q, p) of the phase space. If we choose $-y$ as a control input, $u = -Y(H) = -p/L$, then the energy balance is put in the form

$$\frac{dH}{dt} = -\left(\frac{p}{L}\right)^2 \leq 0. \tag{3.33}$$

This means that the initial Hamiltonian H serves as a Lyapunov function, showing that the origin is an asymptotically stable equilibrium. Once the control is chosen as $u = -p/L = -\dot{q}$, the initial circuit equation becomes

$$L\frac{d^2q}{dt^2} + \frac{dq}{dt} + \frac{1}{C}q = 0,$$

which means that the control serves as a resistance term.

According to [73], a (generalized) port-Hamiltonian system with structure matrix $J(x)$, input matrix $g(x)$, and Hamiltonian H is defined to be

$$\frac{dx}{dt} = J(x)\frac{\partial H}{\partial x} + g(x)f, \quad x \in \mathcal{X}, \quad f \in \mathbb{R}^m,$$
$$e = g^T(x)\frac{\partial H}{\partial x}, \quad e \in \mathbb{R}^m,$$

where $J(x)$ is an $n \times n$ matrix with entries depending smoothly on x, which is assumed to be skew-symmetric, and $x = (x_1, \cdots, x_n)$ are local coordinates for an n-dimensional state space manifold $\mathcal{X}$ (not necessarily even-dimensional). A further theme such as Dirac structure, which is related to port-Hamiltonian systems, will be touched on in Appendix 5.10.3.

Though port-Hamiltonian (or port-controlled Hamiltonian) systems are usually formulated on the phase space $\mathbb{R}^n \times \mathbb{R}^n$ as in [19, 66, 67], the formulation can be set up in the form of (3.32) on a symplectic manifold like a cotangent bundle. A model system for the falling cat will be defined indeed as a port-Hamiltonian system on the cotangent bundle over the shape space in the next chapter.

3.4 Remarks on Optimal Hamiltonians

Aside from port-Hamiltonians, optimal Hamiltonians are introduced through Pontryagin's Maximum Principle in the control theory. The plate-ball problem [8] gives an illustrative example. According to V. Jurdjevic [45], it is formulated as an optimal-control problem on the Lie group $G = \mathbb{R}^2 \times SO(3)$ as follows: The initial setting of the problem is put in the form

$$\frac{dx_1}{dt} = u_1, \quad \frac{dx_2}{dt} = u_2, \quad \frac{dh}{dt} = h\big(-u_1 R(\boldsymbol{e}_2) + u_2 R(\boldsymbol{e}_1)\big), \tag{3.34}$$

where $(x_1, x_2) \in \mathbb{R}^2$, $h \in SO(3)$, and where u_1, u_2 are control functions. We have to note that the control inputs u_1, u_2 do not have the dimension of force or torque and that Eq. (3.34) describes the non-slipping condition for the ball of radius one. Further, Eq. (3.34) is expressed, in terms of left-invariant vector fields, $E_k = (\boldsymbol{e}_k, 0)$, $X_k = (0, hR(\boldsymbol{e}_k))$, $k = 1, 2$, on $G = \mathbb{R}^2 \times SO(3)$, as

$$\frac{dg}{dt} = u_1(E_1 - X_2) + u_2(E_2 + X_1), \quad g \in G. \tag{3.35}$$

The problem is to minimize $\frac{1}{2}\int_0^T (u_1^2 + u_2^2)dt$ among all possible curves $(x_1(t), x_2(t))$ that steer the system from a given initial position and an initial orientation to a prescribed final position and a final orientation. Let (x, h, ξ, Γ) denotes the coordinates of the cotangent bundle, $T^*G = \mathbb{R}^2 \times SO(3) \times \mathbb{R}^2 \times \mathfrak{so}(3)$, of G, where $\xi = \sum_{i=1}^2 \xi_i \boldsymbol{e}_i$ and $\Gamma = \sum_{j=1}^3 \Gamma_j R(\boldsymbol{e}_j)$. Then, on account of (3.35) and the cost function, the Maximum Principle is applied to the function

$$\mathcal{H} = u_1(\xi_1 - \Gamma_2) + u_2(\xi_2 + \Gamma_1) + \frac{\psi_0}{2}(u_1^2 + u_2^2), \tag{3.36}$$

where ψ_0 is a non-positive constant. For $\psi_0 < 0$, one obtains the optimal Hamiltonian

$$H_{\mathrm{op}} = \frac{1}{2}(\xi_1 - \Gamma_2)^2 + \frac{1}{2}(\xi_2 + \Gamma_1)^2, \tag{3.37}$$

where the multiplier has been set at $\psi_0 = -1$. Furthermore, owing to the translation symmetry $\mathbb{R}^2$, the present Hamiltonian system reduces to the Hamiltonian system on $T^*SO(3)$ with the reduced Hamiltonian H_μ obtained by replacing $\mu_k \in \mathbb{R}$ for ξ_k in $H_{\rm op}$. The reduced equations of motion are given by

$$\frac{dh}{dt} = h\Omega_\mu, \quad \frac{d\Gamma}{dt} = [\Gamma, \Omega_\mu], \quad \Omega_\mu := (\mu_2 + \Gamma_1)R(\boldsymbol{e}_1) - (\mu_1 - \Gamma_2)R(\boldsymbol{e}_2). \tag{3.38}$$

If we introduce vectors $\boldsymbol{\gamma}$ and $\boldsymbol{\omega}_\mu$ through $R(\boldsymbol{\gamma}) = \Gamma$ and $R(\boldsymbol{\omega}_\mu) = \Omega_\mu$, Eq. (3.38) is put into

$$\frac{dh}{dt} = hR(\boldsymbol{\omega}_\mu), \quad \frac{d\boldsymbol{\gamma}}{dt} + \boldsymbol{\omega}_\mu \times \boldsymbol{\gamma} = 0, \quad \boldsymbol{\omega}_\mu = \mathrm{grad} H_\mu, \tag{3.39}$$

which are of the same form as (2.28)[30].

Remarks on the falling cat formulation as an optimal-control problem will be touched on in Sect. 4.1 and given in an explicit form in Sect. 4.7.

Chapter 4
The Falling Cat

On the basis of the geometry, mechanics, and control studied in the previous chapters, the falling cat is modeled and analyzed to make the cat model turn a somersault under a well-designed control input.

4.1 Modeling of the Falling Cat

In the previous sections, we have set up mechanics for rigid bodies and (deformable or non-rigid) many-particle systems. In order to deal with the falling cat problem, we need to build up a model of a cat. A way to make up such a model is to extend point particle systems to rigid body systems. This extension can be done in such a manner that the system of point particles is decomposed into segments of which particles are fixed relative to one another. Then, one can extend the geometric setting for point particles to that for rigid bodies and further set up mechanics for rigid bodies so that the constraint of the vanishing total angular momentum may be brought into equations of motion in a skillful manner. A simple cat model of this type is the Kane–Scher model [46] which consists of two identical axial symmetric cylinders jointed by a special type of joint.

Montgomery [62] set up the falling cat problem as a variational problem in the sub-Riemannian geometry on the Kane–Scher model (see Sect. 4.7 for the definition of the sub-Riemannian geodesic problem). While, in the Kane–Scher model [46], the twist-free condition is imposed on the jointed cylinders, we are to study the mechanics and control of the jointed cylinders by relaxing the twist-free condition in order to make the jointed cylinders turn a somersault like a falling cat. Because of the additional degree-of-freedom, the new cat model can make a wide variety of turns.

A further comment is needed on the formulation of the falling cat problem. In [62], the problem is formulated in the sub-Riemannian geometry, in which the

T. Iwai, *Geometry, Mechanics, and Control in Action for the Falling Cat*,
Lecture Notes in Mathematics 2289, https://doi.org/10.1007/978-981-16-0688-5_4

dimension of control variables is not that of torque but of angular velocity. The same formulation is adopted also in [20]. In contrast with this, the control variables have the dimension of torque in our setting. Further, the falling cat model in our setting is treated as a port-controlled Hamiltonian (or port-Hamiltonian for short) system on the cotangent bundle of the shape space for the jointed cylinders. It is to be noted here that the total angular momentum is conserved even if controls are applied as torques acting on momentum variables conjugate to shape coordinates. As is well recognized in control theory, passivity-based control is a well-established technique to design robust controllers for physical systems described by Euler–Lagrange equations of motion. This technique is also described in the Hamiltonian formalism (see [19, 67], and reference therein). The main technique is to stabilize the system by energy-shaping [66]. In the case of the falling cat control, the energy shaping is performed by adding a potential function of shape variables. However, the present potential function is rather artificial and used only for determining a control input. It is obscure that the cat adopts the present technique to design a control for somersault, but the present model is sure to turn a somersault to approach a target state in equilibrium with an expected rotation after finishing a vibrational motion [39].

4.2 Geometric Setting for Rigid Body Systems

As was touched on in Sect. 4.1, the geometric setting for a system of particles can be extended to that for a system of rigid bodies. Suppose we have rigid bodies B_k labeled by k, $k = 1, 2, \ldots, b$. We also assume that the rigid bodies do not collide with one another. Let $\boldsymbol{r}_k$ denote the center-of-mass of B_k. We assume that the standard basis of $\mathbb{R}^3$ coincides with the orthonormal frame with respect to which the inertial tensor A_k of the body B_k takes the form of a diagonal matrix. Then, A_k is put in the diagonal form

$$A_k = \mathrm{diag}(I_{k,1}, I_{k,2}, I_{k,3}),$$

where $I_{k,j}$, $j = 1, 2, 3$, are eigenvalues of A_k. As the configuration of B_k is determined by $(\boldsymbol{r}_k, g_k) \in \mathbb{R}^3 \times SO(3)$, each point $\boldsymbol{x}_{k,a}$ of the rigid body B_k at the generic position is expressed as

$$\boldsymbol{x}_{k,a} = \boldsymbol{r}_k + g_k \boldsymbol{X}_{k,a}, \quad g_k \in SO(3). \tag{4.1}$$

Let M_k denote the mass of the rigid body B_k. Since $\boldsymbol{r}_k$ is the center-of-mass of B_k, $\boldsymbol{X}_{k,a}$ are subject to

$$\sum_a m_{k,a} \boldsymbol{X}_{k,a} = 0, \quad M_k = \sum_a m_{k,a}.$$

The inertia tensor of the rigid body B_k is then defined and arranged as

$$\begin{aligned} A_{k,x}(\boldsymbol{v}) &= \sum_a m_{k,a}(\boldsymbol{r}_k + g_k \boldsymbol{X}_{k,a}) \times \big(\boldsymbol{v} \times (\boldsymbol{r}_k + g_k \boldsymbol{X}_{k,a})\big) \\ &= M_k \boldsymbol{r}_k \times (\boldsymbol{v} \times \boldsymbol{r}_k) + g_k \sum_a m_{k,a} \boldsymbol{X}_{k,a} \times (g_k^{-1} \boldsymbol{v} \times \boldsymbol{X}_{k,a}). \end{aligned} \tag{4.2}$$

Rewriting the second term of the right-hand side of the above equation in terms of the inertia tensor A_k of the rigid body B_k, we have

$$A_{k,x}(\boldsymbol{v}) = M_k \boldsymbol{r}_k \times (\boldsymbol{v} \times \boldsymbol{r}_k) + g_k A_k g_k^{-1} \boldsymbol{v}. \tag{4.3}$$

The total inertia tensor of the rigid body system is given by the sum

$$A_x = \sum_k A_{k,x}. \tag{4.4}$$

If each body B_k has a positive-definite inertia tensor, the total inertia tensor is clearly positive-definite symmetric and thereby invertible.

The total angular momentum of the rigid body system can be defined in a similar manner. For the rigid body B_k, the angular momentum $\boldsymbol{L}_k$ is defined and arranged as

$$\begin{aligned} \boldsymbol{L}_k &= \sum_a m_{k,a}(\boldsymbol{r}_k + g_k \boldsymbol{X}_{k,a}) \times (\dot{\boldsymbol{r}}_k + \dot{g}_k \boldsymbol{X}_{k,a}) \\ &= M_k \boldsymbol{r}_k \times \dot{\boldsymbol{r}}_k + g_k \sum_a m_{k,a} \boldsymbol{X}_a \times (\boldsymbol{\Omega}_k \times \boldsymbol{X}_{k,a}), \end{aligned} \tag{4.5}$$

where $R(\boldsymbol{\Omega}_k) = g_k^{-1} \dot{g}_k$. On using the inertial tensor A_k, $\boldsymbol{L}_k$ is rewritten as

$$\boldsymbol{L}_k = M_k \boldsymbol{r}_k \times \dot{\boldsymbol{r}}_k + g_k A_k \boldsymbol{\Omega}_k. \tag{4.6}$$

The total angular momentum of the rigid body system is the sum of $\boldsymbol{L}_k$,

$$\boldsymbol{L} = \sum_k \boldsymbol{L}_k. \tag{4.7}$$

On expressing the total angular momentum in differential form,

$$\boldsymbol{\Lambda} = \sum_k \boldsymbol{\Lambda}_k, \quad \boldsymbol{\Lambda}_k = M_k \boldsymbol{r}_k \times d\boldsymbol{r}_k + g_k A_k R^{-1}(g_k^{-1} dg_k), \tag{4.8}$$

the connection form ω is defined to be

$$\omega = R(A_x^{-1} \boldsymbol{\Lambda}). \tag{4.9}$$

The present connection form is a natural generalization of the connection form (1.143) for point particles.

The metric for the rigid body system is naturally defined as follows: The metric attached to the rigid body B_k is defined and arranged as

$$\begin{aligned} ds_k^2 &= \sum_a m_a(d\boldsymbol{r}_k + dg_k\boldsymbol{X}_a)\cdot(d\boldsymbol{r}_k + dg_k\boldsymbol{X}_a) \\ &= M_k d\boldsymbol{r}_k\cdot d\boldsymbol{r}_k + \boldsymbol{\Psi}_k\cdot\Big(\sum_a m_a\boldsymbol{X}_a\times(\boldsymbol{\Psi}_k\times\boldsymbol{X}_a)\Big), \end{aligned} \tag{4.10}$$

where $R(\boldsymbol{\Psi}_k) = g_k^{-1}dg_k$. This metric is further rewritten as

$$ds_k^2 = M_k d\boldsymbol{r}_k\cdot d\boldsymbol{r}_k + \boldsymbol{\Psi}_k\cdot A_k\boldsymbol{\Psi}_k, \tag{4.11}$$

where the first and the second terms of the right-hand side of the above equation are associated, respectively, with the kinetic energy of the center-of-mass of the rigid body B_k and the rotational energy of the rigid body B_k about the center-of-mass. The metric for the rigid body system is then given by

$$ds^2 = \sum_k ds_k^2. \tag{4.12}$$

We did not discuss the configuration space of the rigid body system, nor the bundle structure, but have defined the inertia tensor and the connection form in a formal manner. In this geometric setting, the mechanics of these systems can be built up in a similar manner to that in Chap. 2. We do not pursue it, but will discuss the configuration space of the two jointed cylinders in detail and set up mechanics for them in what follows. See also [59], in which rigid body systems are called complexes of rigid molecules.

4.3 Geometric Setting for Two Jointed Cylinders

In this section, geometric settings such as a connection and a metric are defined on the configuration space for two jointed cylinders (Fig. 4.1).

4.3.1 The Configuration Space

A model of the falling cat to be discussed below is composed of two identical axial symmetric cylinders which are jointed together by a special type of joint that will give no constraints on the relative motion of the cylinders other than that

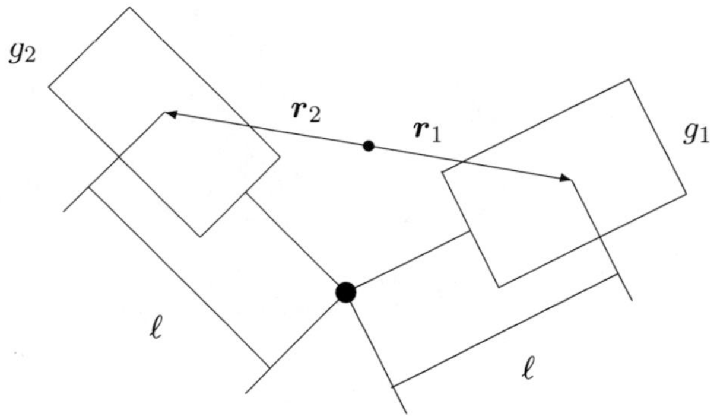

Fig. 4.1 A sketch of two jointed cylinders

they are jointed. Since the center-of-mass of this system is fixed at the origin of $\mathbb{R}^3$, the position vectors, $\boldsymbol{r}_1$ and $\boldsymbol{r}_2$, of the center-of-mass of respective cylinders are determined by the attitude of the cylinders. In fact, letting ℓ be the distance between the joint and the center-of-mass of each cylinder, and g_1 and g_2 rotation matrices describing the attitude of respective cylinders, one has $\boldsymbol{r}_1 + \boldsymbol{r}_2 = 0$ and $\boldsymbol{r}_1 - \boldsymbol{r}_2 = \ell(g_1\boldsymbol{e}_3 - g_2\boldsymbol{e}_3)$, where we have used the fact that the respective cylinders are identical and where $\boldsymbol{e}_a$, $a = 1, 2, 3$, are the standard basis of $\mathbb{R}^3$. Hence, the configuration space X_0 for this system is realized in $\mathbb{R}^3 \times \mathbb{R}^3 \times SO(3) \times SO(3)$ as

$$X_0 = \left\{ \left(\frac{\ell}{2}(g_1\boldsymbol{e}_3 - g_2\boldsymbol{e}_3), -\frac{\ell}{2}(g_1\boldsymbol{e}_3 - g_2\boldsymbol{e}_3), g_1, g_2 \right) \mid (g_1, g_2) \in SO(3) \times SO(3) \right\}, \tag{4.13}$$

which is diffeomorphic to $SO(3) \times SO(3)$ [62].

The rotation group $SO(3)$ acts on X_0 in the manner

$$(\boldsymbol{r}_1, \boldsymbol{r}_2, g_1, g_2) \mapsto (h\boldsymbol{r}_1, h\boldsymbol{r}_2, hg_1, hg_2),$$

where

$$h \in SO(3), \quad \boldsymbol{r}_1 = -\boldsymbol{r}_2 = \frac{\ell}{2}(g_1\boldsymbol{e}_3 - g_2\boldsymbol{e}_3), \quad (g_1, g_2) \in SO(3) \times SO(3).$$

The projection π from X_0 to the factor space or the shape space $M := X_0/SO(3)$ is given by

$$\pi : X_0 \to M \cong SO(3); \quad \pi(\boldsymbol{r}_1, \boldsymbol{r}_2, g_1, g_2) = g_1^{-1} g_2. \tag{4.14}$$

Proposition 4.3.1 *The configuration space X_0 for the jointed cylinders is diffeomorphic with $SO(3) \times SO(3)$ and realized in $\mathbb{R}^3 \times \mathbb{R}^3 \times SO(3) \times SO(3)$ as in* (4.13). *The X_0 has the fiber bundle structure* (4.14) *with structure group $SO(3)$.*

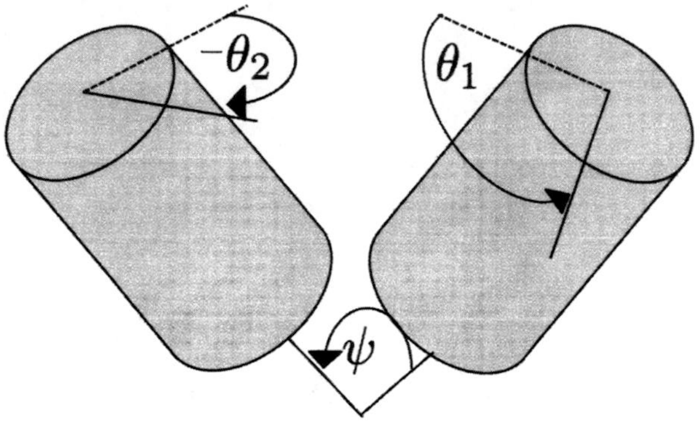

Fig. 4.2 A realization of the local section $\sigma(q)$

We take $(\psi, \theta_1, \theta_2)$ as shape coordinates in M after [62] (see Fig. 4.2). The θ_1 and θ_2 are, respectively, rotation angles about major axes of the respective body halves. The ψ is the angle made by the body halves. Let U be the open subset of M assigned by

$$0 < \psi < \pi, \quad 0 < \theta_1 < 2\pi, \quad 0 < \theta_2 < 2\pi. \tag{4.15}$$

For $q = e^{-\theta_1 R(e_3)} e^{\psi R(e_1)} e^{-\theta_2 R(e_3)} \in U \subset SO(3) \cong M$, we define a local section $\sigma : U \to X_0$ to be $\sigma(q) = (\boldsymbol{\sigma}_1(q), \boldsymbol{\sigma}_2(q), k_1(q), k_2(q))$ with

$$\begin{cases} \boldsymbol{\sigma}_1(q) = \boldsymbol{e}_2 \ell \sin \frac{\psi}{2}, \\ \boldsymbol{\sigma}_2(q) = -\boldsymbol{e}_2 \ell \sin \frac{\psi}{2}, \end{cases} \qquad \begin{cases} k_1(q) = e^{-\frac{\psi}{2} R(e_1)} e^{\theta_1 R(e_3)}, \\ k_2(q) = e^{\frac{\psi}{2} R(e_1)} e^{-\theta_2 R(e_3)}. \end{cases} \tag{4.16}$$

We note here that if $\psi = 0, \pi$, the coordinate system $(\psi, \theta_1, \theta_2)$ fails to work. In fact, according to whether $\psi = 0$ or $\psi = \pi$, the set of coordinate pairs (θ_1, θ_2) and $(\theta_1 + \gamma, \theta_2 - \gamma)$ or that of the pairs (θ_1, θ_2) and $(\theta_1 + \gamma, \theta_2 + \gamma)$ assigns the same shape for an arbitrary constant γ on account of the identities $e^{-\gamma R(e_3)} e^{\gamma R(e_3)} = I$ and $e^{-\gamma R(e_3)} e^{\pi R(e_1)} e^{-\gamma R(e_3)} = e^{\pi R(e_1)}$. However, if the restriction $0 < \psi < \pi$ is imposed, one may take the ranges of θ_1 and θ_2 as $0 \leq \theta_1 \leq 2\pi$ and $0 \leq \theta_2 \leq 2\pi$, respectively. We note in addition that $\boldsymbol{\sigma}_i(q)$, $i = 1, 2$, are determined by $k_i(q)$, $i = 1, 2$:

$$\boldsymbol{\sigma}_1(q) = -\boldsymbol{\sigma}_2(q) = \frac{\ell}{2}(k_1(q)\boldsymbol{e}_3 - k_2(q)\boldsymbol{e}_3).$$

The section σ given in (4.16) is realized by the two cylinders as follows: The jointed cylinders B_1 and B_2 are first set in a manner such that the major axes of respective cylinders are laid along the $\boldsymbol{e}_3$ axis, and then rotated around the $\boldsymbol{e}_3$ axis by the angles θ_1 and $-\theta_2$, respectively, before being turned around the $\boldsymbol{e}_1$ axis by the angles $-\psi/2$

and $\psi/2$, respectively. In general, each point of $\pi^{-1}(U) \cong SO(3) \times U$ is put in the form

$$g\sigma(q) = (g\sigma_1(q), g\sigma_2(q), gk_1(q), gk_2(q)), \quad g \in SO(3). \tag{4.17}$$

We now introduce local coordinates of the structure group $SO(3)$ through

$$g = e^{\chi_2 R(\boldsymbol{e}_2)} e^{\chi_3 R(\boldsymbol{e}_3)} e^{\chi_1 R(\boldsymbol{e}_1)}, \tag{4.18}$$

where we have to notice that the angle variables χ_a, $a = 1, 2, 3$, are taken in a different manner from that for the usual Euler angles. We may take the ranges of χ_a as

$$0 \le \chi_1 \le 2\pi, \quad 0 \le \chi_2 \le 2\pi, \quad -\pi/2 < \chi_3 < \pi/2, \tag{4.19}$$

where the ranges of χ_1 and χ_2 may be shifted as long as the 2π-period is preserved (see (1.90)). A straightforward calculation provides the components of the left-invariant one-form $g^{-1}dg = \sum_a \Psi^a R(\boldsymbol{e}_a)$ as

$$\Psi^1 = d\chi_1 + \sin\chi_3 d\chi_2, \tag{4.20a}$$

$$\Psi^2 = \cos\chi_3 \cos\chi_1 d\chi_2 + \sin\chi_1 d\chi_3, \tag{4.20b}$$

$$\Psi^3 = -\cos\chi_3 \sin\chi_1 d\chi_2 + \cos\chi_1 d\chi_3. \tag{4.20c}$$

The dual vector fields K_a to Ψ^a are given by

$$K_1 = \frac{\partial}{\partial\chi_1}, \tag{4.21a}$$

$$K_2 = \frac{\cos\chi_1}{\cos\chi_3}\frac{\partial}{\partial\chi_2} + \sin\chi_1 \frac{\partial}{\partial\chi_3} - \tan\chi_3 \cos\chi_1 \frac{\partial}{\partial\chi_1}, \tag{4.21b}$$

$$K_3 = -\frac{\sin\chi_1}{\cos\chi_3}\frac{\partial}{\partial\chi_2} + \cos\chi_1 \frac{\partial}{\partial\chi_3} + \tan\chi_3 \sin\chi_1 \frac{\partial}{\partial\chi_1}. \tag{4.21c}$$

We have to note here that though the Ψ^a and the K_a given above have different expressions from (1.95) and (1.98), the Ψ^a and the K_a of the present form play the same role as (1.95) and (1.98), respectively.

4.3.2 Geometric Quantities

We now proceed to write out the inertia tensor, the connection form, and the metric in terms of the local coordinates defined above. We note here that since the cylinders in question are identical and axial symmetric, the masses are equal and the inertia tensor of each cylinder (see (2.11)) takes the form

$$m := M_1 = M_2, \tag{4.22}$$

$$A_1 = A_2 = \mathrm{diag}(I_1, I_1, \nu I_1), \quad \nu := I_3/I_1, \tag{4.23}$$

respectively. We further introduce a parameter λ through

$$m\ell^2 = \lambda I_1. \tag{4.24}$$

From (4.3) and (4.4), the total inertia tensor is defined to be

$$A_x(\boldsymbol{v}) = 2m\boldsymbol{r} \times (\boldsymbol{v} \times \boldsymbol{r}) + \sum_{i=1}^{2} g_i A_i g_i^{-1} \boldsymbol{v}, \tag{4.25}$$

where $\boldsymbol{r} := \boldsymbol{r}_1 = -\boldsymbol{r}_2$, and $\boldsymbol{v} \in \mathbb{R}^3$. The first term of the right-hand side of the above equation stands for the inertia tensor for the collection of the center-of-mass of the cylinders and the second term is the sum of the respective inertia tensors of the cylinders. We note that the symmetry axes of the cylinders are rotated by $k_i(q)$ at the reference position $\sigma(q)$, so that g_i in the above definition are expressed as $g_i = gk_i(q)$, $i = 1, 2$. In particular, the total inertia tensor $A_{\sigma(q)}$ at $\sigma(q) \in X_0$ is expressed as

$$A_{\sigma(q)}(\boldsymbol{v}) = 2m\boldsymbol{\sigma}(q) \times (\boldsymbol{v} \times \boldsymbol{\sigma}(q)) + \sum_{i=1}^{2} k_i(q) A_i k_i^{-1}(q) \boldsymbol{v},$$

where $\boldsymbol{\sigma} := \boldsymbol{\sigma}_1 = -\boldsymbol{\sigma}_2$. The matrix expression of $A_{\sigma(q)}$ is easily found to be

$$A_{\sigma(q)} = 2I_1 \begin{pmatrix} 1 + \lambda \sin^2 \frac{\psi}{2} & 0 & 0 \\ 0 & \cos^2 \frac{\psi}{2} + \nu \sin^2 \frac{\psi}{2} & 0 \\ 0 & 0 & \sin^2 \frac{\psi}{2} + \nu \cos^2 \frac{\psi}{2} + \lambda \sin^2 \frac{\psi}{2} \end{pmatrix}. \tag{4.26}$$

The inertia tensor at a generic point $x = g\sigma(q) \in X_0$ takes the form $A_{g\sigma(q)} = gA_{\sigma(q)}g^{-1}$.

The total angular momentum $\boldsymbol{\Lambda}_x$ as a differential form is defined, from (4.8), to be

$$\boldsymbol{\Lambda}_x = 2m\boldsymbol{r} \times d\boldsymbol{r} + \sum_{i=1}^{2} g_i A_i R^{-1}(g_i^{-1} dg_i). \tag{4.27}$$

In particular, in view of $g_i = gk_i(q)$, the total angular momentum $\mathbf{\Lambda}_{\sigma(q)}$ at $\sigma(q)$ proves to be written out as

$$\begin{aligned}\mathbf{\Lambda}_{\sigma(q)} &= 2m\boldsymbol{\sigma}(q) \times d\boldsymbol{\sigma}(q) + \sum_{i=1}^{2} k_i(q)A_i R^{-1}(k_i(q)^{-1}dk_i(q)) \\ &= \nu I_1\Big(\sin\frac{\psi}{2}(d\theta_1 + d\theta_2)\boldsymbol{e}_2 + \cos\frac{\psi}{2}(d\theta_1 - d\theta_2)\boldsymbol{e}_3\Big). \end{aligned} \tag{4.28}$$

The total angular momentum at $x = g\sigma(q)$ is put in the form

$$\mathbf{\Lambda}_{g\sigma(q)} = g\big(A_{\sigma(q)}(\mathbf{\Psi}) + \mathbf{\Lambda}_{\sigma(q)}\big), \quad R(\mathbf{\Psi}) = g^{-1}dg. \tag{4.29}$$

From (4.25) and (4.27), the connection form is defined to be

$$\omega_x = R(A_x^{-1}(\mathbf{\Lambda}_x)). \tag{4.30}$$

In particular, from (4.26) and (4.28), the connection form $\omega_{\sigma(q)}$ at $\sigma(q)$ turns out to be expressed as

$$\begin{aligned}\omega_{\sigma(q)} &= R(A_{\sigma(q)}^{-1}\mathbf{\Lambda}_{\sigma(q)}) \\ &= \frac{\nu\sin\frac{\psi}{2}}{\cos^2\frac{\psi}{2} + \nu\sin^2\frac{\psi}{2}}\frac{1}{2}(d\theta_1 + d\theta_2)R(\boldsymbol{e}_2) \\ &\quad + \frac{\nu\cos\frac{\psi}{2}}{\sin^2\frac{\psi}{2} + \nu\cos^2\frac{\psi}{2} + \lambda\sin^2\frac{\psi}{2}}\frac{1}{2}(d\theta_1 - d\theta_2)R(\boldsymbol{e}_3). \end{aligned} \tag{4.31}$$

In view of the expression (4.31), we introduce new variables by

$$\phi_1 = \frac{1}{2}(\theta_1 + \theta_2), \quad \phi_2 = \frac{1}{2}(\theta_1 - \theta_2). \tag{4.32}$$

Then, the local coordinates on U we will use below are $(q^i) = (\psi, \phi_1, \phi_2)$. The non-vanishing components of $\omega_{\sigma(q)}$ given in (4.31) are now expressed as

$$\Lambda_2^2(q) = \frac{\nu\sin\frac{\psi}{2}}{\cos^2\frac{\psi}{2} + \nu\sin^2\frac{\psi}{2}}, \quad \Lambda_3^3(q) = \frac{\nu\cos\frac{\psi}{2}}{\sin^2\frac{\psi}{2} + \nu\cos^2\frac{\psi}{2} + \lambda\sin^2\frac{\psi}{2}}, \tag{4.33}$$

and the others vanish. From (4.29) and (4.30), the connection form $\omega_{g\sigma(q)}$ at $g\sigma(q)$ is expressed as

$$\omega_{g\sigma(q)} = \mathrm{Ad}_g(g^{-1}dg + \omega_{\sigma(q)}) = \sum_a \omega^a R(g\boldsymbol{e}_a), \tag{4.34a}$$

$$\omega^a := \Psi^a + \sum_i \Lambda_i^a(q)dq^i. \tag{4.34b}$$

We note here that the one-forms ω^a corresponds to Θ^a given in (1.190).

A vector field Z on X_0 is called horizontal or vibrational if it satisfies $\omega(Z) = 0$. A local basis of the set of horizontal vector fields, which we denote by $(\partial/\partial q^i)^*$, is given by

$$\begin{aligned} \left(\frac{\partial}{\partial\psi}\right)^* &= \frac{\partial}{\partial\psi}, \\ \left(\frac{\partial}{\partial\phi_1}\right)^* &= \frac{\partial}{\partial\phi_1} - \Lambda_2^2(q)K_2, \\ \left(\frac{\partial}{\partial\phi_2}\right)^* &= \frac{\partial}{\partial\phi_2} - \Lambda_3^3(q)K_3, \end{aligned} \tag{4.35}$$

where K_a are given in (4.21). The commutation relations among $(\partial/\partial q^i)^*$ give rise to infinitesimal rotations, which are described as

$$\left[\left(\frac{\partial}{\partial\psi}\right)^*, \left(\frac{\partial}{\partial\phi_1}\right)^*\right] = -\kappa_{12}^2 K_2, \tag{4.36a}$$

$$\left[\left(\frac{\partial}{\partial\psi}\right)^*, \left(\frac{\partial}{\partial\phi_2}\right)^*\right] = -\kappa_{13}^3 K_3, \tag{4.36b}$$

$$\left[\left(\frac{\partial}{\partial\phi_1}\right)^*, \left(\frac{\partial}{\partial\phi_2}\right)^*\right] = -\kappa_{23}^1 K_1, \tag{4.36c}$$

where $\kappa_{12}^2, \kappa_{13}^3, \kappa_{23}^1$ are non-vanishing components of the curvature κ_{ij}^c,

$$\kappa_{ij}^c := \frac{\partial\Lambda_j^c}{\partial q^i} - \frac{\partial\Lambda_i^c}{\partial q^j} - \sum_{a,b} \varepsilon_{abc}\Lambda_i^a\Lambda_j^b, \quad i, k = 1, 2, 3, \tag{4.37}$$

but we do not need to write them out (see also (1.194)). What is worth pointing out is that infinitesimal vibrations are coupled to give rise to infinitesimal rotations. We further note that ω^a and dq^i form a local basis of the space of one-forms, which are dual to the local basis K_a and $(\partial/\partial q^i)^*$ of the space of vector fields;

$$\omega^a(K_b) = \delta_b^a, \quad dq^i(K_b) = 0, \quad \omega^a\left(\left(\frac{\partial}{\partial q^i}\right)^*\right) = 0, \quad dq^i\left(\left(\frac{\partial}{\partial q^j}\right)^*\right) = \delta_j^i. \tag{4.38}$$

Here we make a remark on the range of the variables (ψ, ϕ_1, ϕ_2). On account of $\theta_1 = \phi_1 + \phi_2, \theta_2 = \phi_1 - \phi_2$ and of the periodicity of θ_1 and θ_2, in the (ϕ_1, ϕ_2)-plane, (ϕ_1, ϕ_2) is equivalent to $(\phi_1 + \pi, \phi_2 + \pi)$ and $(\phi_1 + \pi, \phi_2 - \pi)$. Then, the open subset determined by (4.15) is designated by

$$0 < \psi < \pi, \quad 0 < \phi_1 < 2\pi, \quad 0 < \phi_2 < \pi. \tag{4.39}$$

As was remarked in the paragraph after (4.16), if the restriction $0 < \psi < \pi$ is imposed, then one may take the ranges of θ_1 and θ_2 as $0 \le \theta_1 \le 2\pi$ and $0 \le \theta_2 \le 2\pi$, respectively. Correspondingly, the ranges of ϕ_1 and ϕ_2 are chosen to be $0 \le \phi_1 \le 2\pi$ and $0 \le \phi_2 \le \pi$, respectively. Thus, the region determined by $0 \le \phi_1 \le 2\pi$ and $0 \le \phi_2 \le \pi$ is considered as a fundamental region. Further, let us be reminded of the fact that according to whether $\psi = 0$ or $\psi = \pi$, the coordinate pairs (θ_1, θ_2) and $(\theta_1 + \gamma, \theta_2 - \gamma)$ or the pairs (θ_1, θ_2) and $(\theta_1 + \gamma, \theta_2 + \gamma)$ assign the same shape for an arbitrary constant γ. Accordingly, it follows that if $\psi = 0$ then the pairs (ϕ_1, ϕ_2) and $(\phi_1, \phi_2 + \gamma)$ are equivalent for an arbitrary constant γ, and that if $\psi = \pi$, then the pairs (ϕ_1, ϕ_2) and $(\phi_1 + \gamma, \phi_2)$ are equivalent as well.

From (4.11) and (4.12), the metric on X_0 is defined and written out as

$$\begin{aligned} ds^2 &= 2m d\boldsymbol{r} \cdot d\boldsymbol{r} + \sum_{j=1}^{2} \boldsymbol{\Psi}_j \cdot A_j \boldsymbol{\Psi}_j \\ &= \Big(\frac{m}{2}\ell^2 \cos^2\frac{\psi}{2} + \frac{1}{2} I_1\Big)(d\psi)^2 + \Big(2m\ell^2 \sin^2\frac{\psi}{2} + 2I_1\Big)(\Psi^1)^2 \\ &\quad + 2\Big(I_1 \cos^2\frac{\psi}{2} + I_3 \sin^2\frac{\psi}{2}\Big)(\Psi^2)^2 \\ &\quad + \Big(2m\ell^2 \sin^2\frac{\psi}{2} + 2I_1 \sin^2\frac{\psi}{2} + 2I_3 \cos^2\frac{\psi}{2}\Big)(\Psi^3)^2 \\ &\quad + 4I_3 \sin\frac{\psi}{2}\,\Psi^2 d\phi_1 + 4I_3 \cos\frac{\psi}{2}\,\Psi^3 d\phi_2 + 2I_3 (d\phi_1)^2 + 2I_3 (d\phi_2)^2, \end{aligned} \tag{4.40}$$

where $\boldsymbol{r} = \boldsymbol{r}_1 = -\boldsymbol{r}_2$, and where $\boldsymbol{\Psi}_j$ are determined through

$$k_j^{-1} dk_j = R(\boldsymbol{\Psi}_j), \quad j = 1, 2. \tag{4.41}$$

Note that the metric is $SO(3)$ invariant, $ds^2_{\sigma(q)} = ds^2_{g\sigma(q)}$. In terms of dq^i and ω^a with $(q^i) = (\psi, \phi_1, \phi_2)$, the metric is put in the form

$$\begin{aligned} ds^2 &= \frac{1}{2} I_1 \Big(1 + \lambda \cos^2\frac{\psi}{2}\Big)(d\psi)^2 + 2I_1 \frac{\nu \cos^2\frac{\psi}{2}}{\cos^2\frac{\psi}{2} + \nu \sin^2\frac{\psi}{2}}(d\phi_1)^2 \\ &\quad + 2I_1 \frac{(1+\lambda)\nu \sin^2\frac{\psi}{2}}{\sin^2\frac{\psi}{2} + \nu \cos^2\frac{\psi}{2} + \lambda \sin^2\frac{\psi}{2}}(d\phi_2)^2 \end{aligned}$$

$$+2I_1\Big(1+\lambda\sin^2\frac{\psi}{2}\Big)(\omega^1)^2+2I_1\Big(\cos^2\frac{\psi}{2}+\nu\sin^2\frac{\psi}{2}\Big)(\omega^2)^2$$
$$+2I_1\Big(\sin^2\frac{\psi}{2}+\nu\cos^2\frac{\psi}{2}+\lambda\sin^2\frac{\psi}{2}\Big)(\omega^3)^2, \tag{4.42}$$

which we describe compactly as

$$ds^2=\sum_{i,j}a_{ij}dq^i dq^j+\sum_{a,b}A_{ab}\omega^a\omega^b, \tag{4.43}$$

where the quantities A_{ab} denote the components of the inertia tensor (4.26). The quantities (a_{ij}) in the right-hand side of the above equation define a metric on the shape space $M = X_0/SO(3)$. We note that the metric (a_{ij}) is non-degenerate for $0 < \psi < \pi$.

4.3.3 Summary and a Remark on the Geometric Setting

Proposition 4.3.2 *Let* (ψ, ϕ_1, ϕ_2) *be local coordinates of the shape space* M, *where* ϕ_1, ϕ_2 *are given in* (4.32) *with* $q = e^{-\theta_1 R(e_3)}e^{\psi R(e_1)}e^{-\theta_2 R(e_3)} \in SO(3) \cong M$, *and* (χ_1, χ_2, χ_3) *local coordinates of the structure group* $SO(3)$, *which are given in* (4.18). *Then, the connection form* $\omega_{g\sigma(q)}$ *and the metric* ds^2 *on the configuration space* X_0 *are expressed as in* (4.34) *together with* (4.20) *and* (4.33) *and as in* (4.42), *respectively. In particular, the metric projected on* M *is given by the first three terms of the right-hand side of* (4.42).

We have defined the connection form to be $\omega_x = R(A_x^{-1}\mathbf{\Lambda}_x)$ in (4.30) for the two-cylinder system. We now verify that this definition indeed satisfies the conditions

$$\text{(i) } \omega_x(\xi_{X_0}(x)) = \xi, \quad \xi \in \mathfrak{so}(3), \quad \text{(ii) } \omega_{hx} = \mathrm{Ad}_h\omega_x, \quad h \in SO(3), \tag{4.44}$$

where $\xi_{X_0}(x)$ is the fundamental vector field defined on the principal bundle X_0 by

$$\frac{d}{dt}e^{t\xi}x\Big|_{t=0} = \xi_{X_0}(x), \quad x \in X_0, \tag{4.45}$$

and where the symbol $e^{t\xi}x$ in the left-hand side denotes the orbit of x by the action of $e^{t\xi}$. The properties (4.44) are also verified for the connection form on the configuration space for a many-body system (see Proposition 1.7.2 and also Appendix 5.2). We now prove that (4.44) is satisfied. From $\mathbf{\Lambda}_{hx} = h\mathbf{\Lambda}_x$ and

$A_{hx} = hA_x h^{-1}$, we obtain

$$\omega_{hx} = R(A_{hx}^{-1}\mathbf{\Lambda}_{hx}) = \mathrm{Ad}_h R(A_x^{-1}\mathbf{\Lambda}_x) = \mathrm{Ad}_h \omega_x, \tag{4.46}$$

which shows that the condition (ii) holds true. In order to show (i), we need an explicit expression of $\xi_{X_0}(x)$. Since the action of the one-parameter group $e^{t\xi} \in SO(3)$ is expressed as $\boldsymbol{r} \mapsto e^{t\xi}\boldsymbol{r}$ and $g_i \mapsto e^{t\xi}g_i$, the components of $\xi_{X_0}(x)$ are expressed as $\xi\boldsymbol{r}$ and ξg_i. On introducing the vector $\boldsymbol{\theta}$ through $\xi = R(\boldsymbol{\theta})$, $\mathbf{\Lambda}_x(\xi_{X_0}(x))$ is evaluated as

$$\begin{aligned}\mathbf{\Lambda}_x(\xi_{X_0}(x)) &= 2m\boldsymbol{r} \times R(\boldsymbol{\theta})\boldsymbol{r} + \sum_i g_i A_i R^{-1}(g_i^{-1}\xi g_i) \\ &= 2m\boldsymbol{r} \times (\boldsymbol{\theta} \times \boldsymbol{r}) + \sum_i g_i A_i g_i^{-1}\boldsymbol{\theta} \\ &= A_x(\boldsymbol{\theta}).\end{aligned} \tag{4.47}$$

Hence, the connection form is evaluated for ξ_{X_0} as

$$\omega_x(\xi_{X_0}(x)) = R(A_x^{-1}\mathbf{\Lambda}_x(\xi_{X_0}(x))) = R(\boldsymbol{\theta}) = \xi, \tag{4.48}$$

which shows that the condition (i) holds. Thus, Eq. (4.44) is verified.

In addition, we make a comment on the twist-free condition. According to [62], the twist-free condition is described as follows: Let ε be a rotation such that $\varepsilon^2 = I$. Then, an involution i acting on $SO(3) \times SO(3)$ is defined as

$$i(g_1, g_2) = (\varepsilon g_2 \varepsilon, \varepsilon g_1 \varepsilon).$$

The no-twist configuration space is defined to be the subset of fixed points for i, and is shown to be diffeomorphic to $SO(3)$,

$$P_0 = \{(g_1, \varepsilon g_1 \varepsilon);\ g_1 \in SO(3)\} \cong SO(3).$$

At the same time, the structure group reduces to $O(2)$ and the base space to $P_0/O(2) \cong SO(3)/O(2)$, which is diffeomorphic with $S^2/\mathbb{Z}_2 \cong \mathbb{R}P^2$, the real projective space of dimension two. In our model, if $\varepsilon = e^{\pi R(\boldsymbol{e}_2)}$, the twist-free condition for $g_1 = gk_1(q)$ and $g_2 = gk_2(q)$ (see (4.16) and (4.18) for the expression of $k_1(q)$, $k_2(q)$, and g) requires that $\theta_1 = \theta_2$, i.e., $\phi_2 = 0$, and $\chi_1 = \chi_3 = 0$ on account of $\varepsilon R(\boldsymbol{e}_1)\varepsilon = -R(\boldsymbol{e}_1)$, $\varepsilon R(\boldsymbol{e}_2)\varepsilon = R(\boldsymbol{e}_2)$, and $\varepsilon R(\boldsymbol{e}_3)\varepsilon = -R(\boldsymbol{e}_3)$. In accordance with this, the structure group $SO(3)$ reduces to its subgroup $O(2) = \{e^{\chi_2 R(\boldsymbol{e}_2)}\} \cup \{e^{\chi_2 R(\boldsymbol{e}_2)}\}e^{\pi R(\boldsymbol{e}_1)}$. The restriction that $\phi_2 = 0,\ \chi_1 = \chi_3 = 0$ brings the connection form and the metric into

$$\omega^1 = \omega^3 = 0, \quad \omega^2 = d\chi_2 + \frac{\nu \sin\frac{\psi}{2}}{\cos^2\frac{\psi}{2} + \nu \sin^2\frac{\psi}{2}} d\phi_1, \tag{4.49}$$

$$\begin{aligned} ds_0^2 &= \frac{1}{2} I_1 \left(1 + \lambda \cos^2\frac{\psi}{2}\right)(d\psi)^2 + 2I_1 \frac{\nu \cos^2\frac{\psi}{2}}{\cos^2\frac{\psi}{2} + \nu \sin^2\frac{\psi}{2}} (d\phi_1)^2 \\ &+ 2I_1 \left(\cos^2\frac{\psi}{2} + \nu \sin^2\frac{\psi}{2}\right)\left(d\chi_2 + \frac{\nu \sin\frac{\psi}{2}}{\cos^2\frac{\psi}{2} + \nu \sin^2\frac{\psi}{2}} d\phi_1\right)^2, \end{aligned} \tag{4.50}$$

respectively, which have been already obtained in [33, 34] with a slightly different choice of local coordinates.

4.4 A Lagrangian Model of the Falling Cat

We now write out the equations of motion for the jointed cylinders. Let $T(X_0)$ denote the tangent bundle of X_0. Like (2.50) with (2.31), we introduce the variable $\boldsymbol{\pi} = \sum_a \pi^a \boldsymbol{e}_a$, from (4.34b), by

$$\pi^a = \omega^a\left(\frac{d}{dt}\right) = \Psi^a\left(\frac{d}{dt}\right) + \sum \Lambda_i^a(q)\dot{q}^i, \tag{4.51}$$

where $\Psi^a(d/dt)$ denote the quantities Ψ^a, given in (4.20), with $d\chi^a$ replaced with $\dot{\chi}^a$, owing to $d\chi^a(d/dt) = \dot{\chi}^a$, and take $(q, g, \dot{q}, \boldsymbol{\pi})$ as local coordinates of $T(X_0)$. In terms of these new coordinates, the Lagrangian associated with the metric (4.43) takes the form

$$L(q, \dot{q}, \boldsymbol{\pi}) = \frac{1}{2}\sum_{i,j} a_{ij}\dot{q}^i\dot{q}^j + \frac{1}{2}\sum_{a,b} A_{ab}\pi^a\pi^b, \tag{4.52}$$

and is written out as

$$\begin{aligned} L &= \frac{1}{4} I_1 \left(1 + \lambda \cos^2\frac{\psi}{2}\right)(\dot{\psi})^2 + I_1 \frac{\nu \cos^2\frac{\psi}{2}}{\cos^2\frac{\psi}{2} + \nu \sin^2\frac{\psi}{2}} (\dot{\phi}_1)^2 \\ &+ I_1 \frac{(1+\lambda)\nu \sin^2\frac{\psi}{2}}{\sin^2\frac{\psi}{2} + \nu \cos^2\frac{\psi}{2} + \lambda \sin^2\frac{\psi}{2}} (\dot{\phi}_2)^2 \\ &+ I_1 \left(1 + \lambda \sin^2\frac{\psi}{2}\right)(\pi^1)^2 + I_1 \left(\cos^2\frac{\psi}{2} + \nu \sin^2\frac{\psi}{2}\right)(\pi^2)^2 \\ &+ I_1 \left(\sin^2\frac{\psi}{2} + \nu \cos^2\frac{\psi}{2} + \lambda \sin^2\frac{\psi}{2}\right)(\pi^3)^2, \end{aligned} \tag{4.53}$$

where the expression (4.42) of the metric has been used. It is to be noted that L is independent of $g \in SO(3)$ and of ϕ_1 and ϕ_2.

The Euler–Lagrange equations for the present $SO(3)$-invariant Lagrangian take the same form as (2.85),

$$\frac{d}{dt}\frac{\partial L}{\partial \dot{q}^i} - \frac{\partial L}{\partial q^i} - \sum_j \frac{\partial L}{\partial \boldsymbol{\pi}} \cdot \boldsymbol{\kappa}_{ij}\dot{q}^j + \frac{\partial L}{\partial \boldsymbol{\pi}} \cdot (\boldsymbol{\pi} \times \boldsymbol{\lambda}_i) = 0, \tag{4.54a}$$

$$\frac{d}{dt}\frac{\partial L}{\partial \boldsymbol{\pi}} - \frac{\partial L}{\partial \boldsymbol{\pi}} \times \boldsymbol{\pi} + \sum_i \frac{\partial L}{\partial \boldsymbol{\pi}} \times \boldsymbol{\lambda}_i \dot{q}^i = 0, \tag{4.54b}$$

where $\boldsymbol{\kappa}_{ij} = \sum_a \kappa_{ij}^a \boldsymbol{e}_a$, $\boldsymbol{\pi} = \sum_a \pi^a \boldsymbol{e}_a$, and $\boldsymbol{\lambda}_i = \sum_a \Lambda_i^a \boldsymbol{e}_a$. We assume here that we can apply the torque

$$u_1(t)d\psi + u_2(t)d\theta_1 + u_3(t)d\theta_2 = v_1(t)d\psi + v_2(t)d\phi_1 + v_3(t)d\phi_2 \tag{4.55}$$

by using some kind of actuator in order to control the shape of the jointed cylinders. Then, the equations of motion (4.54) become

$$\frac{d}{dt}\frac{\partial L}{\partial \dot{q}^i} - \frac{\partial L}{\partial q^i} - \sum_j \frac{\partial L}{\partial \boldsymbol{\pi}} \cdot \boldsymbol{\kappa}_{ij}\dot{q}^j + \frac{\partial L}{\partial \boldsymbol{\pi}} \cdot (\boldsymbol{\pi} \times \boldsymbol{\lambda}_i) = v_i, \tag{4.56a}$$

$$\frac{d}{dt}\frac{\partial L}{\partial \boldsymbol{\pi}} - \frac{\partial L}{\partial \boldsymbol{\pi}} \times \boldsymbol{\pi} + \sum_i \frac{\partial L}{\partial \boldsymbol{\pi}} \times \boldsymbol{\lambda}_i \dot{q}^i = 0. \tag{4.56b}$$

It is to be noted here that Eq. (4.56b) is equivalent to the conservation of the total angular momentum

$$\boldsymbol{L} = g A_{\sigma(q)} \boldsymbol{\pi}, \quad A_{\sigma(q)} = (A_{ab}), \tag{4.57}$$

where we remark for the proof of this equation that $\boldsymbol{L} = \boldsymbol{\Lambda}_{g\sigma(q)}(d/dt)$ and that from (4.29) and (4.34) the $\boldsymbol{\pi}$ is described as $\boldsymbol{\pi} = \boldsymbol{\Psi}(d/dt) + A_{\sigma(q)}^{-1}\boldsymbol{\Lambda}_{\sigma(q)}(d/dt)$ (distinguish $\boldsymbol{\Lambda}_{\sigma(q)}$ from $\Lambda_i^a(q)$). We note also that the controls v_i have nothing to do with (4.56b).

The motion of the falling cat should be made under the constraint of the vanishing total angular momentum. Since $\boldsymbol{L} = 0$ is equivalent to $\boldsymbol{\pi} = 0$ owing to (4.57), we are allowed to impose the condition $\boldsymbol{\pi} = 0$ on the control equations (4.56). It is to be stressed that the introduction of the variable $\boldsymbol{\pi}$ makes it quite simple to treat the constraint of the vanishing total angular momentum. Since $\partial L/\partial \boldsymbol{\pi} = 0$ if $\boldsymbol{\pi} = 0$, Eq. (4.56b) vanishes, so that Eqs. (4.56) become

$$\frac{d}{dt}\frac{\partial L_c}{\partial \dot{q}^i} - \frac{\partial L_c}{\partial q^i} = v_i, \quad L_c := L|_{\boldsymbol{\pi}=0}. \tag{4.58}$$

The present procedure is interpreted as follows: Even if the torques are applied, the total angular momentum $\boldsymbol{L}$ is conserved, and thereby the Lagrangian control system on $T(X_0)$ is reduced to that on $T(M)$ under the constraint $\boldsymbol{L} = 0$. Since $\partial L_c/\partial\phi_\mu = 0,\ \mu = 1, 2$, the above equations together with (4.53) become

$$\frac{d}{dt}\dot{\psi} = \frac{2}{1+\lambda\cos^2\frac{\psi}{2}}\left(\frac{\lambda}{4}\sin\frac{\psi}{2}\cos\frac{\psi}{2}\dot{\psi}^2 - \frac{\nu^2\sin\frac{\psi}{2}\cos\frac{\psi}{2}}{(\cos^2\frac{\psi}{2}+\nu\sin^2\frac{\psi}{2})^2}\dot{\phi}_1^2\right.$$
$$\left. + \frac{(1+\lambda)\nu^2\sin\frac{\psi}{2}\cos^2\frac{\psi}{2}}{(\sin^2\frac{\psi}{2}+\nu\cos^2\frac{\psi}{2}+\lambda\sin^2\frac{\psi}{2})^2}\dot{\phi}_2^2 + \frac{1}{I_1}\upsilon_1\right), \tag{4.59a}$$

$$\frac{d}{dt}\dot{\phi}_1 = \frac{\cos^2\frac{\psi}{2}+\nu\sin^2\frac{\psi}{2}}{2\nu\cos^2\frac{\psi}{2}}\left(\frac{2\nu^2\sin\frac{\psi}{2}\cos\frac{\psi}{2}}{(\cos^2\frac{\psi}{2}+\nu\sin^2\frac{\psi}{2})^2}\dot{\psi}\dot{\phi}_1 + \frac{1}{I_1}\upsilon_2\right), \tag{4.59b}$$

$$\frac{d}{dt}\dot{\phi}_2 = \frac{\sin^2\frac{\psi}{2}+\nu\cos^2\frac{\psi}{2}+\lambda\sin^2\frac{\psi}{2}}{2(1+\lambda)\nu\sin^2\frac{\psi}{2}}$$
$$\times\left(-\frac{2(1+\lambda)\nu^2\sin\frac{\psi}{2}\cos\frac{\psi}{2}}{(\sin^2\frac{\psi}{2}+\nu\cos^2\frac{\psi}{2}+\lambda\sin^2\frac{\psi}{2})^2}\dot{\psi}\dot{\phi}_2 + \frac{1}{I_1}\upsilon_3\right). \tag{4.59c}$$

Proposition 4.4.1 *Under the condition of the vanishing total angular momentum, the equations of motion for the jointed cylinders with torque inputs are given by the above equations in terms of the local coordinates* (ψ, ϕ_1, ϕ_2) *of the shape space* $M \cong SO(3)$, *where* υ_i, $i = 1, 2, 3$, *denote the torques with respect to the angular velocities* $(\dot{\psi}, \dot{\phi}_1, \dot{\phi}_2)$.

If the torques $\upsilon_i(t)$, $i = 1, 2, 3$, are suitably designed, Eqs. (4.59) should be solved to give functions $\psi(t), \phi_1(t), \phi_2(t)$, with which one can integrate the constraint equation $\boldsymbol{\pi} = 0$ (see (4.51)), i.e.,

$$\frac{dg}{dt} = -g\left(\Lambda_2^2(q)\frac{d\phi_1}{dt}R(\boldsymbol{e}_2) + \Lambda_3^3(q)\frac{d\phi_2}{dt}R(\boldsymbol{e}_3)\right) \tag{4.60}$$

to determine the attitude $g(t)$ of the falling cat. The whole motion of the falling cat with the vanishing total angular momentum is then given by $g(t)\sigma(q(t)) \in X_0$, $q(t) = (\psi(t), \phi_1(t), \phi_2(t))$.

4.5 A Port-Controlled Hamiltonian System

So far we have worked with the falling cat model in the Lagrangian formalism with the constraint of the vanishing total angular momentum. We now discuss the model in the Hamiltonian formalism, taking the constraint into account. As is seen from (2.101), for any $SO(3)$-invariant Hamiltonian H, Hamilton's equations of motion (in the vector form) are expressed as

$$\dot{q}^i = \frac{\partial H}{\partial p_i}, \quad \pi^a = \frac{\partial H}{\partial \varpi_a}, \tag{4.61a}$$

$$\dot{p}_i + \frac{\partial H}{\partial q^i} = \boldsymbol{\varpi} \cdot \sum_j \boldsymbol{\kappa}_{ij} \frac{\partial H}{\partial p_j} - \boldsymbol{\varpi} \cdot \Big(\frac{\partial H}{\partial \boldsymbol{\varpi}} \times \boldsymbol{\lambda}_i \Big), \tag{4.61b}$$

$$\dot{\boldsymbol{\varpi}} = \boldsymbol{\varpi} \times \frac{\partial H}{\partial \boldsymbol{\varpi}} - \sum_i \Big(\boldsymbol{\varpi} \times \boldsymbol{\lambda}_i \Big) \frac{\partial H}{\partial p_i}, \tag{4.61c}$$

where $\boldsymbol{\varpi} = \sum \varpi_a \boldsymbol{e}_a = A_{\sigma(q)} \boldsymbol{\pi}$, being denoted by $\boldsymbol{M}$ in Sect. 2.5. A derivation of these equations in the Hamiltonian formalism with the canonical symplectic form on $T^*(X_0)$ is given in Appendix 5.9. The Hamiltonian H associated with the Lagrangian (4.52) is of the form

$$H = \frac{1}{2} \sum_{i,j} a^{ij} p_i p_j + \frac{1}{2} \sum_{a,b} A^{ab} \varpi_a \varpi_b, \tag{4.62}$$

where $(a^{ij}) = (a_{ij})^{-1}$, which is defined when $0 < \psi < \pi$, and where $(A^{ab}) = (A_{ab})^{-1}$. Thus, in the Hamiltonian formalism, the control equations corresponding to (4.56) are described by (4.61) with the controls v_i added to the right-hand side of (4.61b).

The total angular momentum is given by $g\boldsymbol{\varpi}$ (see (4.57) with $\boldsymbol{\varpi} = A_{\sigma(q)} \boldsymbol{\pi}$). The conservation of the total angular momentum $g\boldsymbol{\varpi}$ is equivalent to Eq. (4.61c), and further Eq. (4.61c) is free from the controls. Thus, we can take into account the constraint $\boldsymbol{\varpi} = 0$ in the Hamiltonian equations (4.61) with the controls added. The resultant equation, which is equivalent to (4.58), is expressed as

$$\dot{q}^i = \frac{\partial H_c}{\partial p_i}, \quad \dot{p}_i + \frac{\partial H_c}{\partial q^i} = v_i, \tag{4.63}$$

where $H_c := H|_{\boldsymbol{\varpi}=0}$ is given by

$$H_c = \frac{1}{2} \sum_{i,j} a^{ij} p_i p_j. \tag{4.64}$$

For the same reason as in the Lagrangian formalism, the control system on $T^*(X_0)$ reduces to that on $T^*(M)$ in spite of the torques applied.

We now focus our interest on the reduced control system on $T^*(M)$. The Hamiltonian vector field X_{H_c} associated with H_c is given, as usual, by

$$X_{H_c} = \sum \Big(\frac{\partial H_c}{\partial p_i} \frac{\partial}{\partial q^i} - \frac{\partial H_c}{\partial q^i} \frac{\partial}{\partial p_i} \Big). \tag{4.65}$$

Let X_i denote the vector fields given by

$$X_i = \frac{\partial}{\partial p_i}, \quad i = 1, 2, 3. \tag{4.66}$$

Then, Eq. (4.63) together with system outputs are expressed as port-controlled Hamilton's equations on $T^*(M)$,

$$\frac{d\tilde{p}}{dt} = X_{H_c} + \sum_{i=1}^{3} v_i X_i, \quad \tilde{p} \in T^*(M), \tag{4.67a}$$

$$y_i = X_i(H_c), \tag{4.67b}$$

where y_i are the system outputs. A frequently used technique is energy-shaping, which is explained as follows: Since the Hamiltonian H_c consists of the kinetic energy only, we can make the system Lyapunov stable by adding a potential function V defined on the shape space M, where V is a positive function which takes a minimum at a point. We denote by q_τ the point at which V takes the minimum. For simplicity, we assume that q_τ is a non-degenerate critical point of V. The point q_τ corresponds to the target shape which the shape of the cat model approaches. The target shape we have in mind is one that the cat takes when she lands on the ground after being launched in the air. Of course, we don't think of the gravitational force, and assume that the center-of-mass of the cat model is fixed. Our idea is that the cat's turn should stop eventually with the target shape, so that the target state is $(q_\tau, 0) \in T^*(M)$. Now, our new Hamiltonian is expressed as

$$\overline{H} = H_c + V, \tag{4.68}$$

and port-controlled Hamilton's equations for $\overline{H}$ become

$$\frac{d\tilde{p}}{dt} = X_{\overline{H}} + \sum_{i=1}^{3} \overline{v}_i X_i, \tag{4.69a}$$

$$\overline{y}_i = X_i(\overline{H}). \tag{4.69b}$$

Equations (4.67) and (4.69) are equivalent if and only if

$$X_V + \sum (\overline{v}_i - v_i) X_i = 0, \tag{4.70a}$$

$$\overline{y}_i - y_i = X_i(V), \tag{4.70b}$$

where X_V denotes the Hamiltonian vector field associated with V. Since $V(q)$ depends only on q, the above equations are put in the form

$$-\frac{\partial V}{\partial q^i} + \overline{v}_i - v_i = 0, \quad \overline{y}_i - y_i = \frac{\partial V}{\partial p_i} = 0. \tag{4.71}$$

If we choose the controls $\overline{v}_i$ in the control system (4.69) by setting

$$\overline{v}_i = -\overline{y}_i, \tag{4.72}$$

then the time derivative of the Hamiltonian $\overline{H}$ is evaluated as

$$\frac{d\overline{H}}{dt} = X_{\overline{H}}(\overline{H}) + \sum_{i=1}^{3} \overline{v}_i X_i(\overline{H}) = -\sum_{i=1}^{3} X_i(\overline{H})^2 \leq 0. \tag{4.73}$$

Since $X_i(\overline{H}) = \sum a^{ij} p_j$ and since (a^{ij}) is positive-definite, it turns out that $X_i(\overline{H}) = 0$ if and only if $p = (p_i) = 0$. This means that the Hamiltonian $\overline{H}$ serves as a Lyapunov function. Since the state $(q_\tau, 0) \in T^*(M)$ is a non-degenerate critical point of $\overline{H}$, it is Lyapunov stable.

Along with the control (4.72), Eq. (4.69a) is written out as

$$\frac{dq^i}{dt} = \frac{\partial H_c}{\partial p_i}, \tag{4.74a}$$

$$\frac{dp_i}{dt} = -\frac{\partial H_c}{\partial q^i} - \frac{\partial V}{\partial q^i} - \frac{\partial H_c}{\partial p_i}. \tag{4.74b}$$

Proposition 4.5.1 *As a port-controlled Hamiltonian system, the falling cat model is subject to the equations* (4.74)*, if a potential function V is chosen. The target state $(q_\tau, 0) \in T^*(M)$ is Lyapunov stable if q_τ is a non-degenerate critical point at which V takes a minimum.*

It is to be noted that Eqs. (4.74) are not Hamilton's equations of motion and also that $\overline{H} = H_c + V$ serves as a Lyapunov function.

4.6 Execution of Somersaults

So far we have formulated the falling cat as a port-controlled Hamiltonian system on $T^*(M)$. The equations (4.74) are integrated to give a curve $(q(t), p(t))$ in $T^*(M)$. Once $(q(t), p(t))$ is obtained, plugging the curve $q(t)$ in M into the constraint equation (4.60) results in a differential equation for $g(t)$, which is easily integrated. For a $g(t)$ thus found, we obtain a vibrational motion $g(t)\sigma(q(t)) \in X_0$ of the falling cat. We now wish to make the model turn a somersault $g(t)\sigma(q(t)) \in X_0$ to perform a prescribed rotation.

Since the target state $(q_\tau, 0) \in T^*(M)$ is Lyapunov stable, for a sufficiently large positive constant T, the distance between $q(T)$ and q_τ in M is negligibly small and $p(T)$ nearly vanishes. Then, we can consider $(q(T), p(T))$ as equal to the target state $(q_\tau, 0)$ in effect. If the initial and target shapes, q_0 and q_τ, respectively, of the falling cat are given, a family of solutions $(q(t), p(t))$ to (4.74) are parameterized by initial values, $K := p(0)$, of $p(t)$, which we denote by $(q(t, K), p(t, K))$. If $g(0)$ is fixed, say, at $g(0) = I$, the identity of $SO(3)$, then the solutions to (4.60) have the parameter K as well, which we refer to as $g(t, K)$. The rotation that the falling cat can gain after finishing her vibrational motion is then given by $g(T, K)$, depending on the parameter K. In order to make the cat model turn by the angle π around the $\boldsymbol{e}_2$ axis, for example, we have to determine the parameter K by solving the equation $g(T, K) = e^{\pi\widehat{\boldsymbol{e}}_2}$ for K (for notational simplicity, $\widehat{\boldsymbol{e}}_a$, $a = 1, 2, 3$, are used in place of $R(\boldsymbol{e}_a)$ here and in what follows). For the existence of solutions to this equation, the initial shape q_0 should be chosen suitably in practice, which is not necessarily the same as the target shape q_τ. The equation for K can be solved by using Newton's method. In fact, if we find an approximate solution K_1 to $g(T, K) = e^{\pi\widehat{\boldsymbol{e}}_2}$, then we can form a sequence K_n which converges to a desired solution K, by Newton's method. To put this procedure into practice, we have to produce a family of numerical solutions $(q(t, K), p(t, K))$ to (4.74) and that of $g(t, K)$ to (4.60).

The potential function we use as an example is expressed as

$$V(q) = k_1 \sin^2\left(\frac{\phi_1 + \phi_2}{2}\right) + k_2 \sin^2\left(\frac{\phi_1 - \phi_2}{2}\right) + k_3 \sin^2\left(\psi - \frac{4\pi}{5}\right). \tag{4.75}$$

The expression of the present potential function means that the target shape is assumed to be $q_\tau = e^{4\pi\widehat{e}_1/5}$. To give an example of the cat's somersault, we set the parameters $I_1, \alpha, \beta, k_1, k_2, k_3$ to be, respectively,

$$I_1 = \frac{19}{12}, \quad \alpha = \frac{6}{19}, \quad \beta = \frac{48}{19}, \quad k_1 = k_2 = k_3 = 1. \tag{4.76}$$

After a number of trials, we can find an initial shape q_0 and a sufficiently large time T such that $q(T)$ approximates to q_τ. According to the above-mentioned procedure, we can solve numerically the equation $g(T, K) = e^{\pi\widehat{e}_2}$ for K. With the K thus found, we can obtain the resultant curves, $q(t, K)$ and $g(t, K)$. An example of such curves with the parameters (4.76) is shown in Figs. 4.3, 4.4, 4.5, 4.6, 4.7, and 4.8. Though the scales of the vertical axes are different from graph to graph, those graphs show that as time goes on, the shape and the attitude tend to the respective targets;

$$q(t) = e^{-(\phi_1(t)+\phi_2(t))\widehat{e}_3} e^{\psi(t)\widehat{e}_1} e^{-(\phi_1(t)-\phi_2(t))\widehat{e}_3} \to e^{\frac{4\pi}{5}\widehat{e}_1} \quad \text{as } t \to \infty,$$

$$g(t) = e^{\chi_2(t)\widehat{e}_2} e^{\chi_3(t)\widehat{e}_3} e^{\chi_1(t)\widehat{e}_1} \to e^{\pi\widehat{e}_2} \quad \text{as } t \to \infty.$$

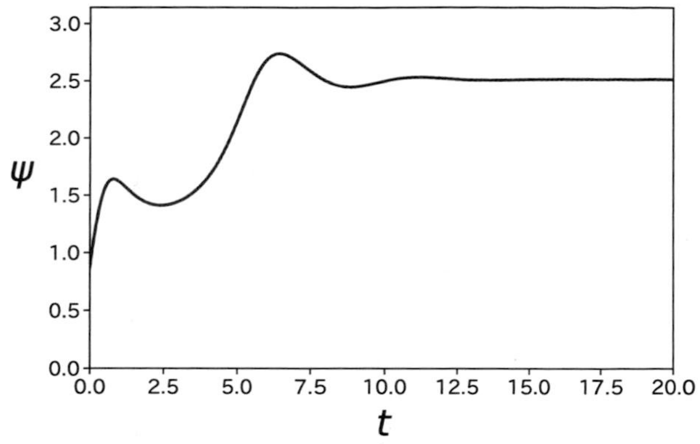

Fig. 4.3 The graph of $\psi(t)$ with $\psi(t) \to 4\pi/5$

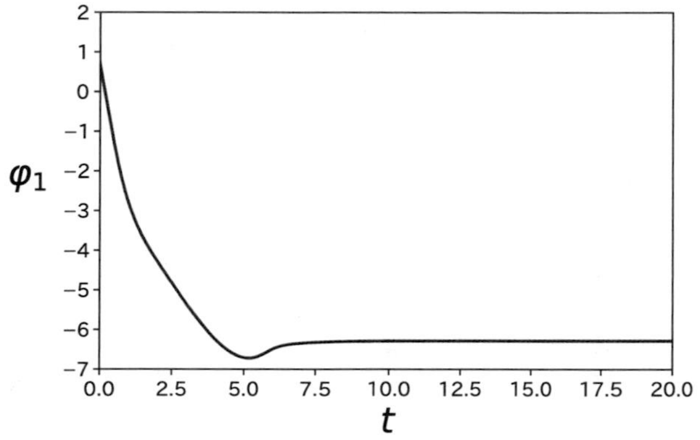

Fig. 4.4 The graph of $\phi_1(t)$ with $\phi_1(t) \to -2\pi$

We can obtain a variety of examples by varying the parameters to observe various somersaults. It is helpful to add some comments on the parameters used in the model. The spring constants k_1 and k_2 in Eq. (4.75) are related to the twist of two cylinders, so that relatively large k_1 and k_2 may indicate a stiffness of the model against twist. If k_1 and k_2 are large, one may imagine that the motion (vibration) that the cat model makes in the air is not swift, so that it takes longer to perform a somersault. In fact, if k_1 and k_2 are sufficiently large, the equation $g(T, K) = e^{\pi \hat{e}_2}$ for K does not have a solution in a region of practical values. This implies that for a successful somersault a flexible body is needed. Cats have more flexible bodies than other animals. This may be a reason why cats can easily turn somersaults.

The numerical data of $q(t)$ and $g(t)$ are used to realize a motion of the jointed cylinders which simulates a cat's somersault. Snapshots of the somersault are given in Fig. 4.9. The series of snapshots starts at the top left (1) and goes zigzag downward to end at the bottom right (10). The time intervals between adjacent snapshots are different from one another. The time interval between the beginning

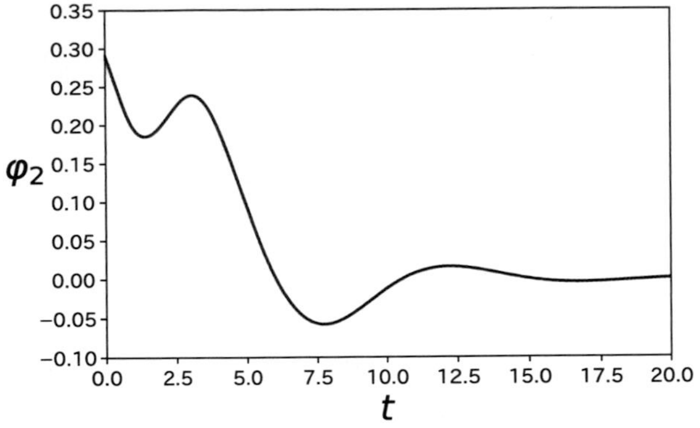

Fig. 4.5 The graph of $\phi_2(t)$ with $\phi_2(t) \to 0$

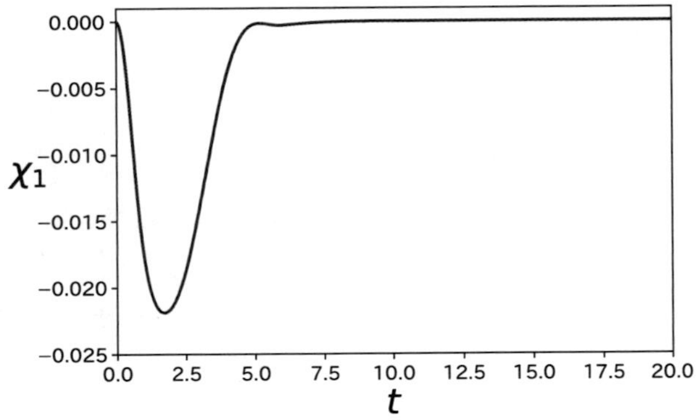

Fig. 4.6 The graph of $\chi_1(t)$ with $\chi_1(t) \to 0$

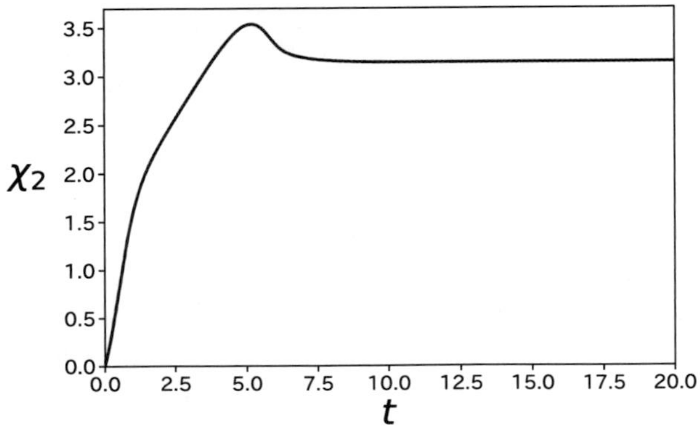

Fig. 4.7 The graph of $\chi_2(t)$ with $\chi_2(t) \to \pi$

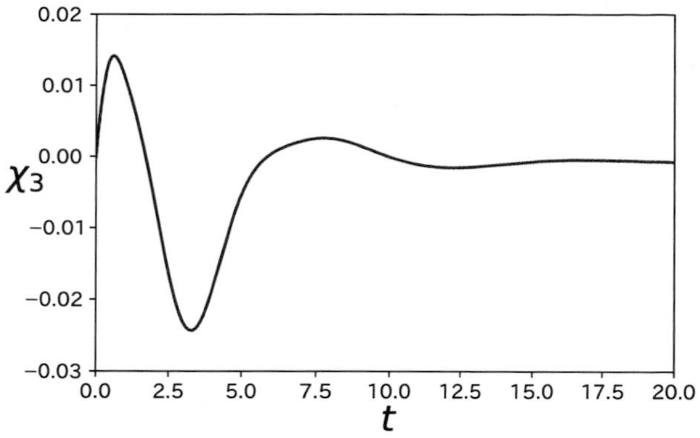

Fig. 4.8 The graph of $\chi_3(t)$ with $\chi_3(t) \to 0$

shots, (1) and (2), is shortest and the succeeding time intervals gradually get larger. Finally, the time interval between the ending snapshots, (9) and (10), becomes largest. A reason why such different time intervals are adopted can be observed from the graphs given in Figs. 4.3, 4.4, 4.5 and in Figs. 4.6, 4.7, 4.8. In fact, almost all changes in the angle variables occur in the beginning half of the whole interval of the time parameter t and few change is made in the latter half of the whole interval. Different time intervals between adjacent snapshots are adopted in order to show visually the amount of change in attitude at each step. A reason why few changes occur in the latter half of the interval is that the target state is asymptotically stable. In this sense, the end figure gives approximately the target attitude of the cat model.

4.7 Remarks on Control Problems

So far we have worked with the falling cat problem on the basis of the port-controlled Hamiltonian system on the cotangent bundle $T^*(M)$. Here we refer to the optimal control of the falling cat using Pontryagin's Maximum Principle [62]. Following the standard procedure, we rewrite the constraint equations (4.60) as

$$\begin{cases} \dfrac{d\psi}{dt} = w^1, \\ \dfrac{d\phi_1}{dt} = w^2, \\ \dfrac{d\phi_2}{dt} = w^3, \end{cases} \qquad \begin{cases} \dfrac{d\chi_1}{dt} = \tan\chi_3\Big(\Lambda_2^2(q)\cos\chi_1\, w^2 - \Lambda_3^3(q)\sin\chi_1\, w^3\Big), \\ \dfrac{d\chi_2}{dt} = \dfrac{1}{\cos\chi_3}\Big(- \Lambda_2^2(q)\cos\chi_1\, w^2 + \Lambda_3^3(q)\sin\chi_1\, w^3\Big), \\ \dfrac{d\chi_3}{dt} = -\Lambda_2^2(q)\sin\chi_1\, w^2 - \Lambda_3^3(q)\cos\chi_1\, w^3, \end{cases} \tag{4.77}$$

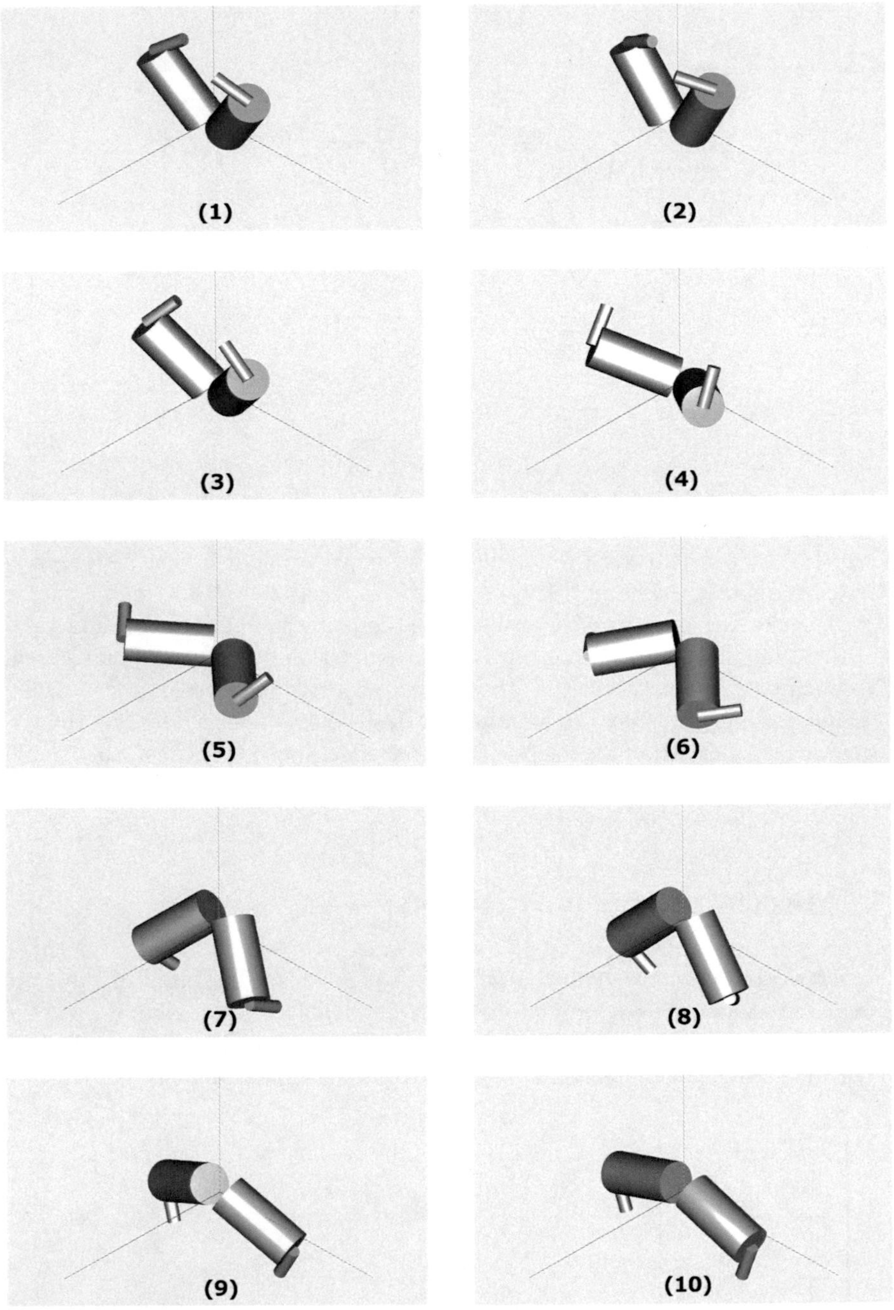

Fig. 4.9 Snapshots of the falling cat turning a somersault, starting at top left (1) and going zigzag downward to end at bottom right (10). Small rods attached to the cylinders are indicators to point out the attitude

and then define a performance index to be

$$\frac{1}{2}\int_0^T \sum_{i,j} a_{ij} w^i w^j dt, \tag{4.78}$$

where the functions w^i denote controls, and where (a_{ij}) is the metric defined on the shape space $M \cong SO(3)$. As for a reason for the choice of this performance index, Montgomery [63] said, "Presumably, she wants to right herself "efficiently", or along a "short path" in shape space." Montgomery [61] defines the sub-Riemannian geodesic problem to be the problem of finding the shortest horizontal curve joining two fixed endpoints $q_0, q_1 \in Q$, where Q is a manifold on which a horizontal distribution is defined together with a smoothly varying positive definite product $\kappa(q)$ on the horizontal subspace of each tangent space $T_q Q$ to Q. This line of study is adopted also in [20] for the Kane–Scher model, and extended for nonholonomic mechanical systems in [50].

Applying the Maximum Principle to

$$\mathcal{H} = \sum_i p_i w^i - \frac{1}{2}\sum_{i,j} a_{ij} w^i w^j, \tag{4.79}$$

we obtain the optimal control $w^i = \sum a^{ij} p_j$ and hence the optimal Hamiltonian,

$$H_{\text{op}} = \frac{1}{2}\sum_{i,j} a^{ij} p_i p_j, \quad (a^{ij}) := (a_{ij})^{-1}. \tag{4.80}$$

Note that $H_{\text{op}} = H_c$ (see (4.64)). Optimal paths will be determined by solving the Hamilton equations associated with H_{op},

$$\dot{q}^i = \frac{\partial H_{\text{op}}}{\partial p_i}, \quad \dot{p}_i = -\frac{\partial H_{\text{op}}}{\partial q^i}. \tag{4.81}$$

From a mechanical point of view, we have to point out that the control inputs w^i adopted in the present procedure are not torques but the angular velocities (see the left half of (4.77)). In this procedure, the controlled bodies are tacitly assumed to be so light that the inertia tensor is negligibly small. In contrast with this, our port-controlled Hamilton equations take the form (4.74), which means that the control inputs are torques indeed.

So far we have thought of the falling cat as being subject to the constraint of the vanishing total angular momentum. This is because the falling cat cannot rotate itself in the air. However, if we take our model as a spacecraft which can rotate itself by some device, then we may assume that we have control inputs other than torque inputs v_i. In this situation, we may define a port-controlled Hamiltonian system on the cotangent bundle $T^*(X_0)$. Since Hamilton's equations of motion without control are given in (4.61), the only task for us to do is to add control inputs to the right-hand sides of (4.61b) and (4.61c). The port-controlled Hamilton equations on $T^*(X_0)$ are then put in a similar form to (4.67),

$$\frac{d\tilde{p}}{dt} = X_H + \sum_{\lambda=1}^{6} v_\lambda X_\lambda, \quad \tilde{p} \in T^*(X_0), \tag{4.82a}$$

$$y_\lambda = X_\lambda(H), \tag{4.82b}$$

where X_λ are vector fields associated with controls; the X_i, $i = 1, 2, 3$, are already given in (4.66), and X_λ, $\lambda = 4, 5, 6$, are defined to be $X_\lambda = \partial/\partial\varpi_a$ with $\lambda = 3+a$. In order to determine control inputs v_λ as functions on $T^*(X_0)$, the energy-shaping can be performed in an analogous manner to that for the port-controlled Hamiltonian system on $T^*(M)$. Like (4.68), we define a new Hamiltonian $\widetilde{H} = H + \widetilde{V}$ on $T^*(X_0)$, where H is the initial Hamiltonian (4.62) and $\widetilde{V}$ is a potential function on X_0. Equations (4.82) are then modified into

$$\frac{d\tilde{p}}{dt} = X_{\widetilde{H}} + \sum_{\lambda=1}^{6} \widetilde{v}_\lambda X_\lambda, \quad \tilde{p} \in T^*(X_0), \tag{4.83a}$$

$$\widetilde{y}_\lambda = X_\lambda(\widetilde{H}). \tag{4.83b}$$

Like (4.72), we choose $\widetilde{v}_\lambda = -\widetilde{y}_\lambda$ as control inputs. By a reasoning similar to (4.70) together with the fact that the Hamiltonian vector field $X_{\widetilde{V}}$ associated to $\widetilde{V}$ is expressed as

$$X_{\widetilde{V}} = \sum_i \Big(-\Big(\frac{\partial}{\partial q^i}\Big)^* \widetilde{V} + \sum_a K_a(\widetilde{V})\Lambda_i^a \Big)\frac{\partial}{\partial p_i} - \sum_a K_a(\widetilde{V})\frac{\partial}{\partial \varpi_a}, \tag{4.84}$$

we eventually find that the control inputs are designed as

$$v_i = -\frac{\partial H}{\partial p_i} - \Big(\frac{\partial}{\partial q^i}\Big)^* \widetilde{V} + \sum_a K_a(\widetilde{V})\Lambda_i^a, \tag{4.85a}$$

$$v_{3+a} = -\frac{\partial H}{\partial \varpi_a} - K_a(\widetilde{V}). \tag{4.85b}$$

We note here that the $X_{\widetilde{V}}$ is obtained by using the method given in Appendix 5.9 with H replaced with $\widetilde{V}$. If the Hamiltonian $\widetilde{H}$ on $T^*(X_0)$ reduces to that on $T^*(M)$, the designed control inputs reduce to $v_i = -\partial H/\partial p_i - \partial V/\partial q^i$, $v_{3+a} = 0$, and accordingly, the controlled Hamilton's equations of motion take the form (4.74).

Chapter 5
Appendices

These appendices contain some remarks on basic notions from differential geometry and advanced materials for many-body systems, together with a review of Newton's law of gravitation.

5.1 Newton's Law of Gravitation, Revisited

Kepler's laws
Kepler's laws of planetary motion are three scientific laws describing the motion of planets around the Sun, which are stated as follows:

(1) The orbit of every planet is an ellipse with the Sun at one of the two foci.
(2) A line segment joining a planet and the Sun sweeps out equal areas during equal intervals of time.
(3) The square of the orbital period of a planet is proportional to the cube of the semi-major axis of its orbit.

The inverse problem with Kepler's laws
Though a falling apple occasioned Newton to research into gravitational law, the universal law of gravity is a consequence of Kepler's laws and Newton's equations of motion. From the viewpoint of differential equations, to find the universal law of gravity is an inverse problem, whereas to solve a given differential equation is a direct problem. The inverse problem we are interested in is the following: Suppose that planetary motion is subject to the equations of motion of the form

$$m\frac{d^2\boldsymbol{r}}{dt^2} = \boldsymbol{F}, \tag{5.1}$$

where the force $\boldsymbol{F}$ is unknown, and that planetary motions provide solutions to these equations and exhibit Kepler's laws. Find the force $\boldsymbol{F}$ that satisfies these conditions.

T. Iwai, *Geometry, Mechanics, and Control in Action for the Falling Cat*,
Lecture Notes in Mathematics 2289, https://doi.org/10.1007/978-981-16-0688-5_5

Consequences of the first and second laws

Kepler's first law says that the solution curves assigned to the planets are planar curves. Let $S(t)$ denote the area swept by the radius vector $\boldsymbol{r}(t)$ during a time interval. As the infinitesimal area swept by the radius vector $\boldsymbol{r}(t)$ for the time interval from t to $t + \Delta t$ is given by

$$\Delta S(t) = \frac{1}{2}|\boldsymbol{r}(t) \times \boldsymbol{r}(t + \Delta t)| = \frac{1}{2}\left|\boldsymbol{r} \times \frac{d\boldsymbol{r}}{dt}\right|\Delta t,$$

the rate of change in $S(t)$ is expressed as

$$\frac{dS}{dt} = \frac{1}{2}\left|\boldsymbol{r} \times \frac{d\boldsymbol{r}}{dt}\right| = \frac{1}{2m}|\boldsymbol{L}|, \tag{5.2}$$

where $\boldsymbol{L}$ denotes the angular momentum defined to be

$$\boldsymbol{L} = \boldsymbol{r} \times m\frac{d\boldsymbol{r}}{dt}. \tag{5.3}$$

Kepler's second law states that $|\boldsymbol{L}|$ is a non-zero constant. Since the trajectory $\boldsymbol{r}(t)$ is sitting on a plane, the angular momentum vector $\boldsymbol{L}$ should be perpendicular to the plane, which means that the orientation of $\boldsymbol{L}$ is also constant. Thus, Kepler's first and second laws are combined to show that the angular momentum $\boldsymbol{L}$ is conserved.

The conservation of the angular momentum together with the equation of motion shows that

$$\frac{d\boldsymbol{L}}{dt} = m\boldsymbol{r} \times \frac{d^2\boldsymbol{r}}{dt^2} = \boldsymbol{r} \times \boldsymbol{F} = 0.$$

This implies that $\boldsymbol{F}$ is proportional to $\boldsymbol{r}$. In other words, $\boldsymbol{F}$ is a central force. Then, we can put the force field in the form

$$\boldsymbol{F} = f(\boldsymbol{r})\frac{\boldsymbol{r}}{r}, \tag{5.4}$$

which means that $\boldsymbol{F}$ is determined by a scalar function $f(\boldsymbol{r})$ only.

In order to determine $f(\boldsymbol{r})$, we reduce the vector equation of motion to a scalar equation of motion including $f(\boldsymbol{r})$. Since the orbit is a planar curve, we take the cylindrical coordinate system (r, θ, z) so that the orbit may be sitting on the x–y plane. We now introduce the moving frame on $\mathbb{R}^3$ by

$$\boldsymbol{e}_r = \begin{pmatrix} \cos\theta \\ \sin\theta \\ 0 \end{pmatrix}, \quad \boldsymbol{e}_\theta = \begin{pmatrix} -\sin\theta \\ \cos\theta \\ 0 \end{pmatrix}, \quad \boldsymbol{e}_z = \begin{pmatrix} 0 \\ 0 \\ 1 \end{pmatrix}, \tag{5.5}$$

which is adapted to the cylindrical coordinate system. Since the orbit is sitting on the x–y plane, the vectors $\{\boldsymbol{e}_r, \boldsymbol{e}_\theta\}$ and $\boldsymbol{e}_z$ are viewed as tangent to and normal to the x–y plane, respectively, when the moving frame is treated in association to the equations of motion.

From the definition, the time derivatives of $\boldsymbol{e}_r$ and $\boldsymbol{e}_\theta$ are easily calculated as

$$\frac{d\boldsymbol{e}_r}{dt} = \boldsymbol{e}_\theta \frac{d\theta}{dt}, \qquad \frac{d\boldsymbol{e}_\theta}{dt} = -\boldsymbol{e}_r \frac{d\theta}{dt}, \tag{5.6}$$

respectively. Then, the time derivative of the radius vector $\boldsymbol{r} = r\boldsymbol{e}_r$ on the x–y plane is shown to take the form

$$\frac{d\boldsymbol{r}}{dt} = \frac{d}{dt}(r\boldsymbol{e}_r) = \frac{dr}{dt}\boldsymbol{e}_r + r\frac{d\theta}{dt}\boldsymbol{e}_\theta. \tag{5.7}$$

By using this equation, the angular momentum becomes

$$\boldsymbol{L} = m\boldsymbol{r} \times \frac{d\boldsymbol{r}}{dt} = mr\boldsymbol{e}_r \times \left(\frac{dr}{dt}\boldsymbol{e}_r + r\frac{d\theta}{dt}\boldsymbol{e}_\theta\right) = mr^2\frac{d\theta}{dt}\boldsymbol{e}_r \times \boldsymbol{e}_\theta = mr^2\frac{d\theta}{dt}\boldsymbol{e}_z,$$

so that one has

$$|\boldsymbol{L}| = L = mr^2\frac{d\theta}{dt}, \tag{5.8}$$

where it has been assumed that $d\theta/dt > 0$. From this, one has $d\theta/dt = L/(mr^2)$. Then, the time derivatives (5.6) of $\boldsymbol{e}_r$ and $\boldsymbol{e}_\theta$ are put in the form

$$\frac{d\boldsymbol{e}_r}{dt} = \frac{L}{mr^2}\boldsymbol{e}_\theta, \qquad \frac{d\boldsymbol{e}_\theta}{dt} = -\frac{L}{mr^2}\boldsymbol{e}_r, \tag{5.9}$$

respectively. Since $d\theta/dt = L/(mr^2)$, Eq. (5.7) is also rewritten as

$$\frac{d\boldsymbol{r}}{dt} = \frac{dr}{dt}\boldsymbol{e}_r + \frac{L}{mr}\boldsymbol{e}_\theta. \tag{5.10}$$

Differentiating further this equation with respect to t, and using the formula (5.9) together with the fact that L is conserved, we obtain

$$\frac{d^2\boldsymbol{r}}{dt^2} = \left(\frac{d^2r}{dt^2} - \frac{L^2}{m^2r^3}\right)\boldsymbol{e}_r. \tag{5.11}$$

This equation and the equation of motion (5.1) with (5.4) are put together to provide the reduced equation of motion,

$$m\left(\frac{d^2r}{dt^2} - \frac{L^2}{m^2r^3}\right) = f(\boldsymbol{r}). \tag{5.12}$$

We again refer to Kepler's first law which states that the orbit is an ellipse. We wish to derive a second-order differential equation to which $r(t)$ is subject along the orbit. As is well known, an ellipse with the semi-major axis a and the semi-minor axis b is expressed, in terms of the planar polar coordinates (r, θ) with one of its foci at the origin, as

$$r = \frac{b^2/a}{1+\varepsilon\cos\theta}, \qquad \varepsilon = \sqrt{1-\frac{b^2}{a^2}}, \tag{5.13}$$

where ε is the eccentricity of the ellipse. Differentiating $(1+\varepsilon\cos\theta)r = b^2/a$ with respect to t, multiplying the resultant equation by r, and using (5.8), we obtain

$$-\varepsilon\sin\theta\frac{d\theta}{dt}r^2 + (1+\varepsilon\cos\theta)r\frac{dr}{dt} = -\varepsilon\sin\theta\frac{L}{m} + \frac{b^2}{a}\frac{dr}{dt} = 0.$$

Further, differentiating this with respect to t, we obtain

$$-\varepsilon\cos\theta\frac{d\theta}{dt}\frac{L}{m} + \frac{b^2}{a}\frac{d^2r}{dt^2} = 0.$$

Multiplying this equation by r^2 and arranging the resultant equation, we obtain

$$\frac{b^2}{a}r^2\frac{d^2r}{dt^2} = \varepsilon r^2\cos\theta\frac{d\theta}{dt}\frac{L}{m} = \varepsilon\cos\theta\frac{L^2}{m^2} = \left(\frac{b^2}{ar}-1\right)\frac{L^2}{m^2}.$$

From this, it follows that the trajectory $r(t)$ along the ellipse is subject to the differential equation

$$\frac{d^2r}{dt^2} = \frac{L^2}{m^2r^3} - \frac{aL^2}{b^2m^2r^2}. \tag{5.14}$$

Comparison of (5.12) and (5.14) results in

$$f(\mathbf{r}) = -\frac{aL^2}{b^2m}\frac{1}{r^2}. \tag{5.15}$$

Though we have obtained the above result on the assumption that the orbits in question are ellipses in accordance with Kepler's first law, we have not imposed the condition $\varepsilon < 1$ for the conic section to be an ellipse, in the manipulation of (5.13) into (5.14). This means that the result (5.15) holds true not only for $\varepsilon < 1$ but also for $\varepsilon = 1,\ \varepsilon > 1$. Put another way, the result (5.15) holds true for all conic sections as orbits of a celestial body.

What the third law means

Equation (5.15) seems to state that the force acting on the planet is an attractive force and its magnitude is proportional to r^{-2}. However, this expression of f depends on the parameters a, b and the constant of motion L, which depends on initial conditions for the motion. We have to express the force independently of such constants, in order to obtain a force law. To this end, we apply Kepler's third law, which is concerned with the orbital period in the case of $\varepsilon < 1$. Then, we have to describe the period in terms of the orbit parameters depending on initial conditions. Since the area swept by the radius vector during the period T is equal to the area enclosed by the ellipse, we have

$$\pi ab = \int_0^T \frac{dS}{dt}dt = \frac{L}{2m}T,$$

where we have used (5.2). Thus, the period is expressed as

$$T = \frac{2m\pi ab}{L}. \tag{5.16}$$

Hence, Kepler's third law is put in the form

$$T = \frac{2m\pi ab}{L} = \gamma a^{3/2}, \quad \gamma = \text{const.},$$

where γ is independent of orbits or takes the same value for all planets. Then, the coefficient of $-r^{-2}$ in the right-hand side of Eq. (5.15) is written as $\frac{aL^2}{b^2m} = \frac{4m\pi^2}{\gamma^2}$, which is a constant independent of orbits. Thus, the force $\boldsymbol{F}$ is found to be described as

$$\boldsymbol{F} = -\frac{4m\pi^2}{\gamma^2}\frac{\boldsymbol{r}}{r^3}.$$

A similar statement to that given in the paragraph after Eq. (5.15) is valid, i.e., the present result holds true even if the orbit of a celestial body is an ellipse, a parabola, or a hyperbola.

The universal law of gravitation

According to Newton's third law (action–reaction law), the Sun is attracted to the planet by a force of the same magnitude. Since the force $\boldsymbol{F}$ is proportional to the mass of the planet, under the symmetric consideration, it should also be proportional to the mass, M, of the Sun. Hence, there exists a universal constant G such that $\frac{4\pi^2}{\gamma^2} = GM$. It then turns out that $\boldsymbol{F}$ is put in the form of the universal law of gravitation,

$$\boldsymbol{F} = -G\frac{mM}{r^2}\frac{\boldsymbol{r}}{r}. \tag{5.17}$$

Once the force of this form is given, the equation of motion (5.1) can be easily integrated to provide conic sections (5.13) as orbits, by using the well-known constants of motion, the angular momentum and the Laplace–Runge–Lenz vector as well as the total energy [21].

5.2 Principal Fiber Bundles

The definition of a principal fiber bundle is given in [48], in which the group action is "to the right". In contrast to this, the group action adopted in the present book is "to the left". The definition given below is modified according to the left action of the group. Let P and M be manifolds and G a Lie group. The P is called a principal fiber bundle over M with G a structure group (or for short, a G-bundle), if the following conditions are satisfied:

(i) G acts on P to the left without fixed point. Put another way, a differentiable map is given

$$G \times P \longrightarrow P, \quad (g, p) \mapsto g \cdot p \tag{5.18}$$

in such a manner that

$$(gh)p = g(hp), \quad g, h \in G, \ p \in P, \tag{5.19a}$$

$$\text{if } a \cdot p = p, \text{ then } a = e, \text{ the identity of } G. \tag{5.19b}$$

(ii) There exists a surjective map (or a projection) $\pi : P \to M$ satisfying

$$\pi(gp) = \pi(p); \tag{5.20a}$$

$$\text{if } \pi(p) = \pi(p'), \text{ then there exists uniquely } g \in G \text{ such that } p' = gp. \tag{5.20b}$$

Put another way, M is the quotient space of P by the equivalent relation induced by G, and the projection π is differentiable.

(iii) P is locally trivial. In other words, for each point x of M, there exists an open subset U such that $\pi^{-1}(U)$ is isomorphic with $U \times G$ in the sense that $p \in \pi^{-1}(U) \mapsto (\pi(p), \phi(p)) \in U \times G$ is a differentiable isomorphism such that $\phi(ap) = a\phi(p)$ for every $a \in G$.

The condition (iii) of the definition of the principal fiber bundle P leads to the existence of an open covering $\{U_\alpha\}$ of M such that

$$p \in \pi^{-1}(U_\alpha) \mapsto (\pi(p), \phi_\alpha(p)) \in U_\alpha \times G$$

is an isomorphism. In this setting, a map $\sigma_\alpha : U_\alpha \to \pi^{-1}(U_\alpha)$ is called a local section if it satisfies

$$\pi(\sigma_\alpha(x)) = x, \quad \phi_\alpha(\sigma_\alpha(x)) = e.$$

Let G_p be the subspace of the tangent space $T_p(Q)$ consisting of vectors tangent to the fiber through p. According to [48] with modification of the group action from right action to left action, a connection in P is an assignment of a subspace Q_p of $T_p(P)$ to each $p \in P$ such that:

(i) $T_p(P) = G_p \oplus Q_p$,
(ii) $Q_{gp} = (L_g)_* Q_p, \quad p \in P, \quad g \in G$, where $L_g p = gp$,
(iii) Q_p depends differentiably on p,
where $(L_g)_*$ denotes the differential map of L_g.

In a dual manner, a connection form ω is defined to be a one-form on P which satisfies:

(i) $\omega(\xi_P) = \xi, \quad \xi \in \mathfrak{g}, \quad \xi_P := \frac{d}{dt} e^{t\xi} g|_{t=0}$,
(ii) $(L_g)^* \omega = \mathrm{Ad}_g \omega$.

If G acts on P to the right, the above conditions (i) and (ii) are replaced by $\xi'_P = \frac{d}{dt} g e^{t\xi}|_{t=0}$ and $(R_g)^* \omega = \mathrm{Ad}_{g^{-1}} \omega$, respectively.

We now make a comment on curvature. According to [48], the curvature form Ω on the principal bundle P is defined to be

$$\Omega(X, Y) = d\omega(X, Y) + \frac{1}{2}[\omega(X), \omega(Y)], \quad X, Y \in T_p(P), \quad p \in P.$$

In contrast to this, the definition given in [64] is expressed as

$$\Omega(X, Y) = d\omega(X, Y) + [\omega(X), \omega(Y)].$$

The definitions given above differ in the second terms of the respective right-hand sides, which is due to the convention of the exterior product of differential forms. Alternatively, the curvature form is expressed as $\Omega = d\omega + \omega \wedge \omega$. However, in this book, the curvature form is defined to be $\Omega = d\omega - \omega \wedge \omega$, as is given in (1.234). The two definitions differ in the respective second terms of the right-hand sides. The difference depends on whether the group action is to the right or to the left.

5.3 Spatial N-Body Systems with $N \geq 4$

The space of $O(3)$ invariants
A way to study the shape space for the spatial N-body system with $N \geq 4$ is to look for invariants under the $SO(3)$ action. To this end, we start by finding invariants

under the $O(3)$ action. Let $X = (\boldsymbol{r}_1, \cdots, \boldsymbol{r}_{N-1}) \in \mathbb{R}^{3\times(N-1)}$ be a matrix consisting of Jacobi vectors. Then, as is easily seen, the matrix

$$Z = X^T X = \begin{pmatrix} |\boldsymbol{r}_1|^2 & \cdots \boldsymbol{r}_1 \cdot \boldsymbol{r}_{N-1} \\ \boldsymbol{r}_2 \cdot \boldsymbol{r}_1 & \cdots \boldsymbol{r}_2 \cdot \boldsymbol{r}_{N-1} \\ \vdots & \vdots \\ \boldsymbol{r}_{N-1} \cdot \boldsymbol{r}_1 & \cdots \ |\boldsymbol{r}_{N-1}|^2 \end{pmatrix}$$

is $O(3)$ invariant. The first thing to note is the fact that rank$X$ = rank$Z$. In fact, since

$$\text{rank}Z = \text{rank}(X^T X) = \dim(\text{Im}(X^T|_{\text{ImX}}) \leq \dim(\text{Im}X) = \text{rank}X,$$

we have the inequality rank$Z \leq$ rank$X$. On the other hand, for $X \in \mathbb{R}^{3\times(N-1)}$ and $Z \in \mathbb{R}^{(N-1)\times(N-1)}$, one has

$$\text{rank}X + \dim(\text{Ker}X) = \text{rank}Z + \dim(\text{Ker}Z) = N - 1.$$

We show that Ker$Z \subset$ Ker$X$. For $\boldsymbol{x} \in$ Ker$Z$, one has $Z\boldsymbol{x} = X^T X\boldsymbol{x} = 0$, which leads to $X\boldsymbol{x} \cdot X\boldsymbol{x} = X^T X\boldsymbol{x} \cdot \boldsymbol{x} = 0$. Thus, one has $X\boldsymbol{x} = 0$, or equivalently, $\boldsymbol{x} \in$ Ker$X$. It then follows that rank$X \leq$ rank$Z$. Thus, we come to the conclusion that rank$X$ = rank$Z$.

The matrix Z is a symmetric and positive-semi-definite matrix, of course. Since X is of rank less than or equal to three, so is Z. Now, we set

$$\mathcal{S} = \{S \in \mathbb{R}^{(N-1)\times(N-1)};\ S^T = S,\ S \geq 0,\ \text{rank}S \leq 3\}, \tag{5.21}$$

where $S \geq 0$ is a symbol meaning that S is positive-semi-definite. The relations among the center-of-mass system Q, the shape space M, and $\mathcal{S}$ are shown by the diagram

$$\begin{array}{ccc} Q & \pi & \\ | & \searrow & \\ | & & M = Q/SO(3), \\ \downarrow & \swarrow & \\ \mathcal{S} & \phi & \end{array} \tag{5.22}$$

where the map $\phi : M \to \mathcal{S}$ is defined to be

$$\phi(\pi(X)) = X^T X. \tag{5.23}$$

Note that the map ϕ is well-defined, since $\phi(\pi(gX)) = X^T X$ for $g \in SO(3)$. The parity operator P on M is defined to be

$$P\pi(X) = \pi(-X), \quad X \in Q.$$

Proposition 5.3.1 *The map $\phi : M \to S$ is a two-sheeted covering, in general. To be precise, if $\phi(\pi(X)) = \phi(\pi(X'))$ then $\pi(X') = P\pi(X)$ or $\pi(X') = \pi(X)$ for X and X' with* rank$X$ = rankX' = 3, *but $\pi(X') = \pi(X)$ for X and X' with $r =$* rank$X$ = rankX' *and $0 \leq r \leq 2$.*

Proof Suppose $\pi(X)$ and $\pi(X')$ have the same target, $\phi(\pi(X)) = \phi(\pi(X'))$. Let rank$X$ = rank$X' = r \leq 3$. Let λ_j, $j = 1, \ldots, N-1$, be the eigenvalues of $X^T X = X'^T X'$ with $\lambda_1 \geq \cdots \geq \lambda_r > \lambda_{r+1} = \cdots = \lambda_{N-1} = 0$, and $\boldsymbol{v}_1, \cdots, \boldsymbol{v}_r, \boldsymbol{v}_{r+1}, \cdots, \boldsymbol{v}_{N-1}$ the orthonormal eigenvectors associated with the eigenvalues λ_j, $j = 1, \ldots, N-1$, respectively. We denote by v_j^α the components of $\boldsymbol{v}_j \in \mathbb{R}^{N-1}$, $\alpha, j = 1, \ldots, N-1$, and define the vectors $\boldsymbol{x}_k \in \mathbb{R}^3$, by using $X = (\boldsymbol{r}_1, \cdots, \boldsymbol{r}_{N-1})$, to be

$$\boldsymbol{x}_k = \sum_{\alpha=1}^{N-1} v_k^\alpha \boldsymbol{r}_\alpha, \quad k = 1, \ldots, N-1.$$

The inner products among these vectors are

$$\boldsymbol{x}_k \cdot \boldsymbol{x}_\ell = \sum_{\alpha,\beta=1}^{N-1} v_k^\alpha v_\ell^\beta \boldsymbol{r}_\alpha \cdot \boldsymbol{r}_\beta = \sum_{\alpha,\beta=1}^{N-1} v_k^\alpha Z_{\alpha\beta} v_\ell^\beta = \sum_{\alpha=1}^{N-1} v_k^\alpha \lambda_\ell v_\ell^\alpha = \lambda_\ell \delta_{k\ell}.$$

In particular, one has $\boldsymbol{x}_\ell \cdot \boldsymbol{x}_\ell = \lambda_\ell$. Since $\lambda_\ell = 0$ for $\ell \geq r+1$, we find that $\boldsymbol{x}_\ell = 0$ for $\ell \geq r+1$. For $k = 1, \ldots, r$, we introduce the vectors $\hat{\boldsymbol{e}}_k = \frac{1}{\sqrt{\lambda_k}} \boldsymbol{x}_k$, which form an orthonormal system in $\mathbb{R}^3$. In a similar manner, by using $\boldsymbol{r}'_\alpha$ from X', we can define another orthonormal system $\hat{\boldsymbol{e}}'_k$, $k = 1, \ldots, r$. For both orthonormal systems, there exists $h \in O(3)$ such that

$$h\hat{\boldsymbol{e}}_k = \hat{\boldsymbol{e}}'_k, \quad k = 1, \ldots, r.$$

If $0 \leq r \leq 2$, then h can be chosen from $SO(3)$, but if $r = 3$, h cannot be restricted to $SO(3)$.

Incidentally, the Jacobi vectors $\boldsymbol{r}_\alpha$ (resp. $\boldsymbol{r}'_\alpha$) can be described in terms of $\hat{\boldsymbol{e}}_k$ (resp. $\hat{\boldsymbol{e}}'_k$). In fact, we define the following vectors and arrange them to obtain

$$\sum_{k=1}^{r} \sqrt{\lambda_k} v_k^\beta \hat{\boldsymbol{e}}_k = \sum_{\alpha=1}^{N-1} \sum_{k=1}^{r} v_k^\beta v_k^\alpha \boldsymbol{r}_\alpha = \sum_{\alpha=1}^{N-1} \sum_{k=1}^{N-1} v_k^\beta v_k^\alpha \boldsymbol{r}_\alpha = \sum_{\alpha=1}^{N-1} \delta_{\beta\alpha} \boldsymbol{r}_\alpha = \boldsymbol{r}_\beta,$$

where we have used the fact that $\boldsymbol{x}_k = 0$ for $k > r$ and the fact that $\{\boldsymbol{v}_j\}$ form an orthonormal system in $\mathbb{R}^{N-1}$, which is alternatively expressed as $V^T V = I$ and $VV^T = I$ in terms of $V = (\boldsymbol{v}_1, \cdots, \boldsymbol{v}_{N-1})$. In the same manner, we obtain $\sum_{k=1}^{r} \sqrt{\lambda_k} v_k^\beta \hat{\boldsymbol{e}}_k' = \boldsymbol{r}_\beta'$. It then follows that

$$h\boldsymbol{r}_\alpha = \boldsymbol{r}_\alpha', \quad \alpha = 1, \ldots, N-1, \quad h \in O(3),$$

or equivalently, $hX = X'$. If $h \in O(3)$ and $h \notin SO(3)$, then $\pi(hX) = \pi(-X) = P\pi(X)$. If $h \in SO(3)$, then $\pi(hX) = \pi(X)$, of course.

The last task to do is to show that ϕ is surjective. Let $S \in \mathcal{S}$ be a positive-semi-definite symmetric matrix of rank less than or equal to three. We denote the eigenvalues of S by $\lambda_1 \geq \cdots \geq \lambda_r > \lambda_{r+1} = \cdots = \lambda_{N-1} = 0$ and the associated orthonormal eigenvectors by $\boldsymbol{u}_1, \cdots, \boldsymbol{u}_r, \boldsymbol{u}_{r+1}, \cdots, \boldsymbol{u}_{N-1}$, respectively. Then, S is expressed as

$$S = (\boldsymbol{u}_1, \cdots, \boldsymbol{u}_r, \boldsymbol{u}_{r+1}, \cdots, \boldsymbol{u}_{N-1})\Lambda(\boldsymbol{u}_1, \cdots, \boldsymbol{u}_r, \boldsymbol{u}_{r+1}, \cdots, \boldsymbol{u}_{N-1})^T,$$

where Λ is the diagonal matrix with entries $\lambda_i \delta_{ij}$, $i, j = 1, \ldots, N-1$. On denoting the components of $\boldsymbol{u}_k$ by u_k^α, the components of S are expressed as

$$S_{\alpha\beta} = \sum_{k=1}^{N-1} u_k^\alpha \lambda_k u_k^\beta = \sum_{k=1}^{r} u_k^\alpha \lambda_k u_k^\beta.$$

We now define the Jacobi vectors to be

$$\boldsymbol{r}_\alpha = \sum_{k=1}^{r} \sqrt{\lambda_k} u_k^\alpha \hat{\boldsymbol{e}}_k, \quad \alpha = 1, \ldots, N-1.$$

Then, the inner products among $\boldsymbol{r}_\alpha$ are calculated as

$$\boldsymbol{r}_\alpha \cdot \boldsymbol{r}_\beta = \sum_{k=1}^{r} u_k^\alpha \lambda_k u_k^\beta = S_{\alpha\beta}.$$

This means that there exists a system of Jacobi vectors $X = (\boldsymbol{r}_1, \cdots, \boldsymbol{r}_{N-1})$ such that $X^T X = S$. Hence, ϕ proves to be surjective. This completes the proof.

The shape space for the four-body system
According to the rank of the configuration of the Jacobi vectors, the center-of-mass system Q is decomposed into

$$Q = \sqcup_{k=0}^{3} Q_k, \quad Q_k := \{X = (\boldsymbol{r}_1, \cdots, \boldsymbol{r}_{N-1}) | \operatorname{rank}(X) = k\}. \tag{5.24}$$

On account of Proposition 5.3.1, the factor spaces $Q_k/SO(3)$ for $k = 0, 1, 2$, are homeomorphic to the subspace, $\mathcal{S}_k$, of $\mathcal{S}$ consisting of positive-semi-definite matrices of rank k, but the factor space $Q_3/SO(3)$ is a double cover of the space, $\mathcal{S}_3$, of positive-definite matrices of rank three. Clearly, $\mathcal{S}_0 = \{0\}$. We now consider the $\mathcal{S}_1$. Any positive-semi-definite symmetric matrix $S \in \mathbb{R}^{(N-1)\times(N-1)}$ of rank one is put in the form $\boldsymbol{v}_1\lambda_1\boldsymbol{v}_1^T$, where λ_1 is a positive eigenvalue of S, and $\boldsymbol{v}_1 \in \mathbb{R}^{N-1}$ is the associated eigenvector. Since $\boldsymbol{v}_1\lambda_1\boldsymbol{v}_1^T = (-\boldsymbol{v}_1)\lambda_1(-\boldsymbol{v}_1)^T$, one has $\mathcal{S}_1 = Q_1/SO(3) \cong \mathbb{R}_+ \times \mathbb{R}P^{N-2}$.

Before attending to $\mathcal{S}_2$ and $\mathcal{S}_3$, we recall the fact that the space of positive-semi-definite symmetric matrices forms a convex cone in the space of symmetric matrices. In fact, for any positive-semi-definite symmetric matrix A and any positive number μ, μA is also a positive-semi-definite symmetric matrix and further for positive-semi-definite symmetric matrices $A, B \in \mathbb{R}^{n\times n}$, one verifies that $\boldsymbol{x} \cdot (\lambda A + (1-\lambda)B)\boldsymbol{x} \geq 0$ for $0 \leq \lambda \leq 1$, $\boldsymbol{x} \in \mathbb{R}^n$, which means that $\lambda A + (1-\lambda)B$ is positive-semi-definite as well.

In what follows, we confine ourselves to the center-of-mass system Q of spatial four bodies. Then, the $\mathcal{S}$ given in (5.21) with $N = 4$ is the set of positive-semi-definite symmetric 3×3 matrices. Hence, $\mathcal{S}$ is a convex cone in the space, $\tilde{\mathcal{S}}$, of 3×3 symmetric matrices, where $\tilde{\mathcal{S}}$ is homeomorphic to $\mathbb{R}^6$. Let

$$\mathcal{B} = \{S \in \tilde{\mathcal{S}};\ \mathrm{tr}S = 1\}.$$

Then, the intersection, $\mathcal{S} \cap \mathcal{B}$, of the convex cone and a hyperplane in $\mathbb{R}^6$ is homeomorphic to $\overline{D}^5$, the five-dimensional closed disk. Thus, $\mathcal{S} - \{0\}$ is homeomorphic to $\mathbb{R}_+ \times \overline{D}^5$. The subspace $\mathbb{R}_+ \times D^5$ corresponds to the set of positive-definite symmetric 3×3 matrices, and the boundary $\mathbb{R}_+ \times S^4$ to the set of non-positive-definite symmetric 3×3 matrices except for $\{0\}$ [65]. It then turns out that the $\mathcal{S}_3$ and the union $\mathcal{S}_1 \cup \mathcal{S}_2$ are homeomorphic to $\mathbb{R}_+ \times D^5$ and to $\mathbb{R}_+ \times S^4$, respectively.

The double of $\mathcal{S}_3$ is homeomorphic to $\mathbb{Z}_2 \times \mathbb{R}_+ \times D^5 \cong \mathbb{R}_+ \times (S^5 - S^4)$, where D^5 is doubled to be the northern and the southern hemispheres of S^5 and where S^4 denotes the equator of S^5. In the boundary of $\mathbb{R}_+ \times (S^5 - S^4)$, $\mathcal{S}_2 \cup \mathcal{S}_1$ is realized so as to be $\mathcal{S}_2 \cong \mathbb{R}_+ \times (S^4 - \mathbb{R}P^2)$ and $\mathcal{S}_1 \cong \mathbb{R}_+ \times \mathbb{R}P^2$. The boundary of $\mathcal{S}_1$ is $\mathcal{S}_0 = \{0\}$. In particular, the subspace of positive-semi-definite symmetric matrices of rank less than or equal to two, $\mathcal{S}_2 \cup \mathcal{S}_1 \cup \mathcal{S}_0$, is the linear space homeomorphic $\mathbb{R}^5$.

Proposition 5.3.2 *The shape space for the spatial four-body system is stratified into*

$$Q_3/SO(3) \cong \mathbb{R}_+ \times (S^5 - S^4),$$

$$Q_2/SO(3) \cong \mathbb{R}_+ \times (S^4 - \mathbb{R}P^2),$$

$$Q_1/SO(3) \cong \mathbb{R}_+ \times \mathbb{R}P^2,$$

$$Q_0/SO(3) \cong \{0\}.$$

In particular, the shape space of non-singular configurations and the total shape space are diffeomorphic, respectively, to

$$\dot{Q}/SO(3) = Q_3/SO(3) \cup Q_2/SO(3) \cong \mathbb{R}_+ \times (S^5 - \mathbb{R}P^2),$$
$$Q/SO(3) = \dot{Q}/SO(3) \cup Q_1/SO(3) \cup Q_0/SO(3) \cong \mathbb{R}^6.$$

Now that we have identified the shape space for the spatial four-body system, we proceed to $SO(3)$-invariants in order to describe the shape space. The six quantities $\boldsymbol{r}_\alpha \cdot \boldsymbol{r}_\beta$ with $1 \leq \alpha \leq \beta \leq 3$ are $O(3)$-invariants. As an $SO(3)$ invariant, we take $\boldsymbol{r}_1 \cdot (\boldsymbol{r}_2 \times \boldsymbol{r}_3)$. According to Littlejohn–Reinsch [52], the following quantities $(v, w_1, \cdots, w_5)$ are adopted as coordinates of the shape space $Q/SO(3) \cong \mathbb{R}^6$,

$$v = \boldsymbol{r}_1 \cdot (\boldsymbol{r}_2 \times \boldsymbol{r}_3), \quad w_1 = \frac{\sqrt{3}}{2}(|\boldsymbol{r}_1|^2 - |\boldsymbol{r}_2|^2), \quad w_2 = \sqrt{3}\boldsymbol{r}_1 \cdot \boldsymbol{r}_2,$$
$$w_3 = \sqrt{3}\boldsymbol{r}_2 \cdot \boldsymbol{r}_3, \quad w_4 = \sqrt{3}\boldsymbol{r}_3 \cdot \boldsymbol{r}_1, \quad w_5 = \frac{1}{2}(-|\boldsymbol{r}_1|^2 - |\boldsymbol{r}_2|^2 + 2|\boldsymbol{r}_3|^2). \tag{5.25}$$

In the rest of this section, we will touch on the shape space of the N-body system in general. We begin with a review of the singular-value decomposition of matrices.

The singular-value decomposition of matrices

Lemma 5.3.1 *For a matrix $A \in \mathbb{R}^{m\times n}$ of rank r, there exist orthogonal matrices $U \in O(m)$ and $V \in O(n)$ such that*

$$A = U\Sigma V^T, \quad \Sigma = \begin{pmatrix} S & 0 \\ 0 & 0 \end{pmatrix} \in \mathbb{R}^{m\times n}, \tag{5.26}$$

where $S = \mathrm{diag}(\sigma_1, \cdots, \sigma_r)$ with $\sigma_1 \geq \sigma_2 \geq \cdots \geq \sigma_r > 0$.

Proof Since $A^T A$ is a positive-semi-definite symmetric matrix, it has non-negative eigenvalues, which we denote by $\sigma_1^2, \cdots, \sigma_n^2$, where $\sigma_j \geq 0$ are called the singular values of A and arranged in descending order,

$$\sigma_1 \geq \sigma_2 \geq \cdots \geq \sigma_r > 0 = \sigma_{r+1} = \cdots = \sigma_n.$$

We denote by $\boldsymbol{v}_1, \cdots, \boldsymbol{v}_r, \boldsymbol{v}_{r+1}, \cdots, \boldsymbol{v}_n$ the normalized eigenvectors associated with σ_j^2, respectively. Here, we note that the latter orthonormal vectors $\boldsymbol{v}_{r+1}, \cdots, \boldsymbol{v}_n$ are subject to freedom of choice. Put another way, any orthonormal vectors $\boldsymbol{v}'_{r+j} = \sum_{k=1}^{n-r} \boldsymbol{v}_{r+k} h_{kj}$ with $(h_{kj}) \in O(n-r)$ can be adopted in place of $\boldsymbol{v}_{r+j}$, $j = 1, \cdots, n-r$. Let

$$V_1 = (\boldsymbol{v}_1, \cdots, \boldsymbol{v}_r), \quad V_2 = (\boldsymbol{v}_{r+1} \cdots, \boldsymbol{v}_n), \quad V = (V_1, V_2),$$

where V_1 and V_2 satisfy

$$V_1^T V_1 = I_r, \quad V_2^T V_2 = I_{n-r},$$

respectively, and where I_r and I_{n-r} denote the $r \times r$ and the $(n-r) \times (n-r)$ identity matrices, respectively. Then, by the definition of eigenvectors, we have

$$A^T A V_1 = V_1 S^2, \quad A^T A V_2 = 0, \quad S = \mathrm{diag}(\sigma_1, \cdots, \sigma_r).$$

By multiplying V_1^T and V_2^T to the left of the first and the second of the above equations, respectively, and by arranging the resulting equations, we obtain

$$(A V_1 S^{-1})^T A V_1 S^{-1} = I_r, \quad (A V_2)^T A V_2 = 0,$$

respectively. The first of the above equations means that $U_1 := A V_1 S^{-1}$ is a collection of orthonormal vectors in $\mathbb{R}^m$, and the second implies that $A V_2 = 0$. Let U_2 be a collection of orthonormal vectors in $\mathbb{R}^m$ such that $U = (U_1, U_2) \in O(m)$, where the choice of U_2 is not unique, and subject to the transformation $O(m-r)$.

Now, by multiplying U^T and V to the left and the right of A, respectively, we obtain

$$U^T A V = \begin{pmatrix} U_1^T A V_1 & U_1^T A V_2 \\ U_2^T A V_1 & U_2^T A V_2 \end{pmatrix} = \begin{pmatrix} S & 0 \\ 0 & 0 \end{pmatrix} = \Sigma,$$

where we have used the fact that

$$A V_1 = U_1 S, \quad A V_2 = 0, \quad U_2^T U_1 = 0.$$

Thus, A is decomposed into $A = U \Sigma V^T$. This completes the proof.

Stratification of the shape space

Before attending to Eq. (5.26), we refer to Steifel manifolds. The space of r-frames in $\mathbb{R}^m$ is called a Stiefel manifold and denoted by $V_r(\mathbb{R}^m)$ or $V(m, r)$ [64], and further realized as $V_r(\mathbb{R}^m) = O(m)/O(m-r) = SO(m)/SO(m-r)$. Eq. (5.26) is rewritten as

$$A = U_1 S V_1^T, \quad U_1^T U_1 = I_r, \quad S > 0, \quad V_1^T V_1 = I_r,$$

where U_1 and V_1 belong to the Stiefel manifolds $V_r(\mathbb{R}^m)$ and $V_r(\mathbb{R}^n)$, respectively, and S is a positive-definite symmetric matrix. However, such a decomposition is not unique. In fact, one obtains another decomposition

$$A = (U_1 h^{-1}) h S h^{-1} (V_1 h^{-1})^T, \quad h \in O(r),$$

where hSh^{-1} is an $r \times r$ positive-definite symmetric matrix, and where U_1h^{-1} and $(V_1h^{-1})^T$ are subject to the conditions $(U_1h^{-1})^T U_1h^{-1} = I_r$ and $(V_1h^{-1})^T(V_1h^{-1}) = I_r$, respectively. It then turns out that in general, any linear map $\phi : \mathbb{R}^n \to \mathbb{R}^m$ of rank r is decomposed into the composition $\phi = \iota \circ \sigma \circ \pi$ of linear maps,

$$\mathbb{R}^n \xrightarrow{\pi} \mathbb{R}^r \xrightarrow{\sigma} \mathbb{R}^r \xrightarrow{\iota} \mathbb{R}^m,$$

where $\pi : \mathbb{R}^n \to \mathbb{R}^r$ and $\iota : \mathbb{R}^r \to \mathbb{R}^m$ are subject to the conditions $\pi \circ \pi^T = \mathrm{id}_{\mathbb{R}^r}$ and $\iota^T \circ \iota = \mathrm{id}_{\mathbb{R}^r}$, respectively, and $\sigma : \mathbb{R}^r \to \mathbb{R}^r$ is a positive-definite symmetric map. We denote by $\mathbb{R}^{m\times n}(r)$ and $S^+(r)$ the set of $m \times n$ real matrices of rank r and the set of $r \times r$ positive-definite real symmetric matrices. Then, the above decomposition of ϕ implies that

$$\mathbb{R}^{m\times n}(r) \cong \frac{V_r(\mathbb{R}^m) \times S^+(r) \times V_r(\mathbb{R}^n)}{O(r)},$$

where the equivalence relation by $O(r)$ is defined to be

$$(\iota, \sigma, \pi) \sim (\iota h^{-1}, h\sigma h^{-1}, h\pi), \quad h \in O(r).$$

We apply the decomposition of $\mathbb{R}^{m\times n}(r)$ in order to stratify the center-of-mass system. The center-of-mass system Q of the spatial N-body system is viewed as the set of $3 \times (N-1)$ real matrices consisting of Jacobi column vectors and stratified according to the rank of such matrices, so that we obtain the stratification

$$\mathbb{R}^{3\times(N-1)} = \bigsqcup_{0\le r\le \min(3,N-1)} \frac{V_r(\mathbb{R}^3) \times S^+(r) \times V_r(\mathbb{R}^{N-1})}{O(r)}.$$

Accordingly, the shape space $M = Q/SO(3)$ is stratified into the disjoint union of the factor spaces of respective strata by $SO(3)$ [80].

$SO(3)$ invariants

For the four-body system, we have already found coordinates (5.25) of the shape space. In the rest of this section, without reference to the detailed topology of the shape space, we will give coordinates of the shape space for the N-body system with $N \geq 4$, which will amount to finding invariants of $SO(3)$. According to (1.119) or (5.24), the shape space is stratified into

$$M = M_0 \cup M_1 \cup \dot{M},$$

where

$$M_k = Q_k/SO(3), \;\; k = 0, 1, 2, 3, \quad \dot{M} = M_2 \cup M_3.$$

The space M_0 is a singleton. We turn to $M_1 = Q_1/SO(3)$. Let $X = (\boldsymbol{r}_1, \cdots, \boldsymbol{r}_{N-1}) \in Q_1$. Since rank$X = 1$, there are numbers ξ_k and a unit vector $\boldsymbol{u}$ such that

$$\boldsymbol{r}_k = \xi_k \boldsymbol{u}, \quad k = 1, \ldots, N-1,$$

where there is a number $0 \leq \ell \leq N-1$ such that $\xi_\ell > 0$ with $\boldsymbol{u}$ being replaced by $-\boldsymbol{u}$ if necessary. Then, X is arranged as

$$X = (\boldsymbol{r}_1, \cdots, \boldsymbol{r}_{N-1}) = \boldsymbol{u}\xi_\ell \Big(\frac{\xi_1}{\xi_\ell}, \cdots, \frac{\xi_{\ell-1}}{\xi_\ell}, 1, \frac{\xi_{\ell+1}}{\xi_\ell}, \ldots, \frac{\xi_{N-1}}{\xi_\ell} \Big),$$

where $\boldsymbol{u} \in \mathbb{R}^{3\times 1}$ and $(\xi_1/\xi_\ell, \cdots, \xi_{N-1}/\xi_\ell) \in \mathbb{R}^{1\times(N-1)}$ and where $\xi_\ell = |\boldsymbol{r}_\ell| > 0$, $\xi_k = (\boldsymbol{r}_k \cdot \boldsymbol{r}_\ell)/|\boldsymbol{r}_\ell|$, which are $SO(3)$ invariant. Thus, we have found the local coordinates $\xi_k/\xi_\ell, k \neq \ell$ of $M_1 \cong \mathbb{R}_+ \times \mathbb{R}P^{N-2}$.

We proceed to find local coordinates on $\dot{M}$. Since rank$(\boldsymbol{r}_1, \cdots, \boldsymbol{r}_{N-1}) \geq 2$, there exist linearly independent vectors $\boldsymbol{r}_i, \boldsymbol{r}_j$ with $i \neq j$. We may take $i = 1, j = 2$ without loss of generality. Then, there exist orthonormal vectors $\boldsymbol{u}_1, \boldsymbol{u}_2$ such that

$$\boldsymbol{r}_1 = \xi_1 \boldsymbol{u}_1, \quad \boldsymbol{r}_2 = \xi_2 \boldsymbol{u}_1 + \xi_3 \boldsymbol{u}_2, \quad \xi_1 > 0,\ \xi_3 > 0. \tag{5.27}$$

We introduce $\boldsymbol{u}_3$ by $\boldsymbol{u}_3 = \boldsymbol{u}_1 \times \boldsymbol{u}_2$. Then, each $\boldsymbol{r}_k$ can be expressed as a linear combination of $\boldsymbol{u}_a, a = 1, 2, 3$, so that there exist a matrix $Y \in \mathbb{R}^{3\times(N-1)}$ such that the configuration X is expressed as

$$X = (\boldsymbol{r}_1, \cdots, \boldsymbol{r}_{N-1}) = (\boldsymbol{u}_1, \boldsymbol{u}_2, \boldsymbol{u}_3)Y, \quad Y = \begin{pmatrix} \xi_1 & \xi_2 & \xi_4 & & \xi_{3N-8} \\ 0 & \xi_3 & \xi_5 & \cdots & \xi_{3N-7} \\ 0 & 0 & \xi_6 & & \xi_{3N-6} \end{pmatrix}, \tag{5.28}$$

where $\xi_j \in \mathbb{R},\ j = 1, 2, \ldots, 3N-6$. As is easily seen, the quantities ξ_j are $SO(3)$ invariant. These quantities can be expressed as functions of the Jacobi vectors, as will be shown below. First, from (5.27), we easily obtain

$$\xi_1 = |\boldsymbol{r}_1|, \quad \xi_2 = \frac{\boldsymbol{r}_1 \cdot \boldsymbol{r}_2}{|\boldsymbol{r}_1|}, \quad \xi_3 = \frac{|\boldsymbol{r}_1 \times \boldsymbol{r}_2|}{|\boldsymbol{r}_1|}.$$

Further calculation shows that the unit vectors $\boldsymbol{u}_k$ are described as

$$\boldsymbol{u}_1 = \frac{\boldsymbol{r}_1}{|\boldsymbol{r}_1|}, \quad \boldsymbol{u}_2 = \frac{\boldsymbol{r}_1 \times (\boldsymbol{r}_2 \times \boldsymbol{r}_1)}{|\boldsymbol{r}_1 \times (\boldsymbol{r}_2 \times \boldsymbol{r}_1)|}, \quad \boldsymbol{u}_3 = \frac{\boldsymbol{r}_1 \times \boldsymbol{r}_2}{|\boldsymbol{r}_1 \times \boldsymbol{r}_2|},$$

respectively. To deal with $\xi_j, j \geq 4$, we refer to the definition of Y, according to which, the Jacobi vectors $\boldsymbol{r}_\ell, 3 \leq \ell \leq N-1$, are expressed as

$$\boldsymbol{r}_\ell = \xi_{3\ell-5}\boldsymbol{u}_1 + \xi_{3\ell-4}\boldsymbol{u}_2 + \xi_{3\ell-3}\boldsymbol{u}_3, \quad 3 \leq \ell \leq N-1.$$

Now it is straightforward to verify that

$$\xi_{3\ell-5} = \frac{\boldsymbol{r}_1 \cdot \boldsymbol{r}_\ell}{|\boldsymbol{r}_1|}, \quad \xi_{3\ell-4} = \frac{(\boldsymbol{r}_\ell \times \boldsymbol{r}_1) \cdot (\boldsymbol{r}_2 \times \boldsymbol{r}_1)}{|\boldsymbol{r}_1||\boldsymbol{r}_2 \times \boldsymbol{r}_1|}, \quad \xi_{3\ell-3} = \frac{(\boldsymbol{r}_1 \times \boldsymbol{r}_2) \cdot \boldsymbol{r}_\ell}{|\boldsymbol{r}_1 \times \boldsymbol{r}_2|}.$$

Our last task is to show that the map

$$\xi : M \to \mathbb{R}^{3N-6}, \quad \pi(X) \longmapsto (\xi_1, \ldots, \xi_{3N-6})$$

is injective. Suppose that $\xi(\pi(X)) = \xi(\pi(X'))$. Then, one has

$$X = (\boldsymbol{r}_1, \cdots, \boldsymbol{r}_{N-1}) = (\boldsymbol{u}_1, \boldsymbol{u}_2, \boldsymbol{u}_3)Y,$$
$$X' = (\boldsymbol{r}'_1, \cdots, \boldsymbol{r}'_{N-1}) = (\boldsymbol{v}_1, \boldsymbol{v}_2, \boldsymbol{v}_3)Y,$$

where Y is the matrix with entries ξ_j, as is given in (5.28), and where $\boldsymbol{u}_k$ and $\boldsymbol{v}_k$ are both orthonormal systems with $\boldsymbol{u}_3 = \boldsymbol{u}_1 \times \boldsymbol{u}_2$ and $\boldsymbol{v}_3 = \boldsymbol{v}_1 \times \boldsymbol{v}_2$. Since the orientation of both orthonormal systems are the same, there exist $g \in SO(3)$ such that

$$(\boldsymbol{v}_1, \boldsymbol{v}_2, \boldsymbol{v}_3) = g(\boldsymbol{u}_1, \boldsymbol{u}_2, \boldsymbol{u}_3).$$

Hence, one obtains $\pi(X) = \pi(X')$, which shows that the map ξ is injective. It then turns out that $(\xi_1, \cdots, \xi_{3N-6})$ serve as local coordinates of the shape space $\dot{M}$.

So far we have studied the shape space for many-body systems with $SO(3)$ action. See [76] for shape spaces under $O(3)$ action. In addition, it is worth mentioning that apart from many-body mechanics, shape spaces form an interesting subject in statistical science [47].

5.4 The Orthogonal Group $O(n)$

Before studying many-body systems in $\mathbb{R}^n$, we present a review of the orthogonal group $O(n)$. The groups defined to be

$$O(n) = \{X \in \mathbb{R}^{n \times n};\ X^T X = I_n\},$$

$$SO(n) = \{X \in \mathbb{R}^{n \times n};\ X^T X = I_n,\ \det X = 1\}$$

are called the orthogonal group and the rotation group, respectively. The condition $X^T X = I_n$ implies that X preserves the inner product on $\mathbb{R}^n$. Put another way, for any $\boldsymbol{x}, \boldsymbol{y} \in \mathbb{R}^n$, the equation $X\boldsymbol{x} \cdot X\boldsymbol{y} = \boldsymbol{x} \cdot \boldsymbol{y}$ holds true, where the dot $\cdot$ indicates the standard inner product on $\mathbb{R}^n$. It is easily seen that both $O(n)$ and $SO(n)$ form groups. These groups are also known as Lie groups, i.e., they are algebraically groups and topologically manifolds. We touch here on the difference between $O(n)$

and $SO(n)$. From $X^T X = I_n$, it follows that det $X = \pm 1$. This implies that $O(n)$ consists of two subsets, one of which is determined by det $X = 1$ and the other by det $X = -1$ and no intersection of them exists. The subset determined by det $X = 1$ is refereed to as $SO(n)$.

We now explain how $O(n)$ can be viewed as a submanifold of $\mathbb{R}^{n\times n} = \mathbb{R}^{n^2}$. Let $X = (x_{ij})$. Then, the defining equations of $O(n)$ are expressed as

$$f_{ij}(X) = \sum_{k=1}^{n} x_{ki}x_{kj} - \delta_{ij} = 0, \quad 1 \le i \le j \le n,$$

which impose $n(n+1)/2$ conditions on $\mathbb{R}^{n\times n}$. We wish to show that the functions f_{ij} are indeed independent of one another. If they are independent, on account of the following proposition (see [15] for proof), $O(n)$ is a submanifold of $\mathbb{R}^{n\times n}$.

Proposition 5.4.1 *Let $f : \mathbb{R}^m \to \mathbb{R}^k$ be a smooth map, and $v_0 \in \mathbb{R}^k$. If $f^{-1}(v_0) \neq \emptyset$ and if $Df(x) : \mathbb{R}^m \to \mathbb{R}^k$ is surjective for all $x \in f^{-1}(v_0)$, then $f^{-1}(v_0)$ is an $(m-k)$-dimensional submanifold of $\mathbb{R}^m$, where $Df(x)$ denotes the derivative of f at x.*

In order to apply Proposition 5.4.1 with $m = n^2$ and $k = n(n+1)/2$, we proceed to the proof of the mutual independence of ∇f_{ij} on $O(n)$, where ∇f_{ij} denotes the gradient of $f_{ij}(X)$. To show the independence of the gradient vectors, we apply a lemma on Grammian matrices:

Lemma 5.4.1 *For r $(r \le m)$ vectors $\boldsymbol{a}_j = (a_{kj})$ and $\boldsymbol{b}_j = (b_{kj})$ of $\mathbb{R}^m$ with $1 \le k \le m$, $1 \le j \le r$, the determinant of the $r \times r$ matrix $(\boldsymbol{a}_i \cdot \boldsymbol{b}_j)$ is given by*

$$\det(\boldsymbol{a}_i \cdot \boldsymbol{b}_j) = \sum_{i_1 < \cdots < i_r} \begin{vmatrix} a_{i_1 1} & \cdots & a_{i_1 r} \\ \vdots & \cdots & \vdots \\ a_{i_r 1} & \cdots & a_{i_r r} \end{vmatrix} \begin{vmatrix} b_{i_1 1} & \cdots & b_{i_1 r} \\ \vdots & \cdots & \vdots \\ b_{i_r 1} & \cdots & b_{i_r r} \end{vmatrix}.$$

Then, for the Grammian matrix $(\boldsymbol{a}_i \cdot \boldsymbol{a}_j)$, one has $\det(\boldsymbol{a}_i \cdot \boldsymbol{a}_j) \ge 0$. This implies that $\det(\boldsymbol{a}_i \cdot \boldsymbol{a}_j) > 0$ if and only if $\boldsymbol{a}_j$, $j = 1, \cdots, r$, are linearly independent. In particular, for $r = m$, one has $\det(\boldsymbol{a}_i \cdot \boldsymbol{a}_j) = (\det A)^2$, where $A = (\boldsymbol{a}_1, \cdots, \boldsymbol{a}_m)$.

This lemma can be easily verified by using the very definition that the determinant of the matrix $U = (\boldsymbol{u}_1, \cdots, \boldsymbol{u}_r)$ with $\boldsymbol{u}_j \in \mathbb{R}^r$ is an alternating multilinear function of the columns $\boldsymbol{u}_j$ that maps the identity matrix to 1, and the formula $\det U = \det U^T$. Then, the $\det(\boldsymbol{u}_1, \cdots, \boldsymbol{u}_r)$ with the column vectors $\boldsymbol{u}_j = (\boldsymbol{a}_1 \cdot \boldsymbol{b}_j, \cdots, \boldsymbol{a}_r \cdot \boldsymbol{b}_j)^T$ is viewed as an alternating multilinear function in $\boldsymbol{a}_i$ and $\boldsymbol{b}_j$.

Now, we put $X, Y \in \mathbb{R}^{n\times n}$ in the form $X = (\boldsymbol{x}_1, \cdots, \boldsymbol{x}_n)$ and $Y = (\boldsymbol{y}_1, \cdots, \boldsymbol{y}_n)$, respectively. Then, the inner product of $X, Y \in \mathbb{R}^{n\times n}$ is described as $\langle X, Y\rangle = \mathrm{tr}(X^T Y) = \sum_k \boldsymbol{x}_k \cdot \boldsymbol{y}_k$. In this notation, the function f_{ij} is expressed as $f_{ij}(X) =$

$\boldsymbol{x}_i \cdot \boldsymbol{x}_j - \delta_{ij}$, and the differential of f_{ij}, $df_{ij} = \langle \nabla f_{ij}, dX \rangle$, is written out as

$$df_{ij} = d\boldsymbol{x}_i \cdot \boldsymbol{x}_j + \boldsymbol{x}_i \cdot d\boldsymbol{x}_j,$$

so that the matrix ∇f_{ij} is put in the form

$$\nabla f_{ij} = (0, \cdots, 0, \underbrace{\boldsymbol{x}_j}_{i}, 0, \cdots, 0, \underbrace{\boldsymbol{x}_i}_{j}, 0, \cdots, 0) \in \mathbb{R}^{n \times n}.$$

We now calculate the inner product of ∇f_{ij} and $\nabla f_{k\ell}$ to obtain

$$\langle \nabla f_{ij}, \nabla f_{k\ell} \rangle = \delta_{ik} \boldsymbol{x}_j \cdot \boldsymbol{x}_\ell + \delta_{j\ell} \boldsymbol{x}_i \cdot \boldsymbol{x}_k + \delta_{i\ell} \boldsymbol{x}_j \cdot \boldsymbol{x}_k + \delta_{jk} \boldsymbol{x}_i \cdot \boldsymbol{x}_\ell.$$

Evaluated on $O(n)$, this inner product becomes

$$\langle \nabla f_{ij}, \nabla f_{k\ell} \rangle \big|_{O(n)} = 2(\delta_{ik}\delta_{j\ell} + \delta_{i\ell}\delta_{jk}),$$

where use has been made of $\boldsymbol{x}_j \cdot \boldsymbol{x}_k = \delta_{jk}$. These are $(ij, k\ell)$ components of the Grammian matrix formed by the gradient vectors ∇f_{ij}. This Grammian matrix is written as a matrix with double indices,

$$\begin{pmatrix} 4I_n & \\ & 2I_{n(n-1)/2} \end{pmatrix},$$

where the first n rows and n columns have double indices with $i = j,\ k = \ell$, and the remaining $n(n-1)/2$ rows and $n(n-1)/2$ columns have double indices with $i < j,\ k < \ell$. Since this matrix is regular on $O(n)$, $\nabla f_{ij},\ 1 \le i \le j \le n$, have been verified to be independent of one another on the whole $O(n)$. This ends the proof.

For $O(n),\ SO(n)$, their Lie algebras are defined to be

$$\mathfrak{o}(n) = \{A \in \mathbb{R}^{n \times n};\ e^{tA} \in O(n),\ t \in \mathbb{R}\},$$

$$\mathfrak{so}(n) = \{A \in \mathbb{R}^{n \times n};\ e^{tA} \in SO(n),\ t \in \mathbb{R}\},$$

respectively. It is an easy matter to show that $\mathfrak{o}(n) = \mathfrak{so}(n)$:

$$\mathfrak{o}(n) = \mathfrak{so}(n) = \{A \in M(n, \mathbb{R});\ A + A^T = 0\}.$$

Clearly, the $\mathfrak{o}(n)$ is a linear space, and closed with respect to the commutator operation, that is, if $A, B \in \mathfrak{o}(n)$, then $[A, B] \in \mathfrak{o}(n)$.

5.5 Many-Body Systems in n Dimensions

The isomorphism $\wedge^2\mathbb{R}^n \cong \mathfrak{so}(n)$

In order to study many-body systems in n dimensions, we need to introduce a generalization of the map $R : \mathbb{R}^3 \to \mathfrak{so}(3)$. We begin by introducing exterior product operation. For $\boldsymbol{x}, \boldsymbol{y} \in \mathbb{R}^n$ and $a \in \mathbb{R}$, the exterior product operation $(\boldsymbol{x}, \boldsymbol{y}) \mapsto \boldsymbol{x} \wedge \boldsymbol{y}$ is defined in such a manner that

$$\boldsymbol{x} \wedge \boldsymbol{y} = -\boldsymbol{y} \wedge \boldsymbol{x}, \quad (\boldsymbol{x}_1 + \boldsymbol{x}_2) \wedge \boldsymbol{y} = \boldsymbol{x}_1 \wedge \boldsymbol{y} + \boldsymbol{x}_2 \wedge \boldsymbol{y}, \quad (a\boldsymbol{x}) \wedge \boldsymbol{y} = a(\boldsymbol{x} \wedge \boldsymbol{y}).$$

Let $\boldsymbol{e}_j$ denote the standard basis of $\mathbb{R}^n$. We denote the totality of the linear combinations of $\boldsymbol{e}_i \wedge \boldsymbol{e}_j$, $i < j$, by

$$\wedge^2\mathbb{R}^n = \Big\{ \sum_{i<j} c_{ji}\boldsymbol{e}_i \wedge \boldsymbol{e}_j : c_{ij} \in \mathbb{R} \Big\},$$

which forms a vector space with the basis $\boldsymbol{e}_i \wedge \boldsymbol{e}_j$, $i < j$. For given coefficients $c_{ji}, i < j$, we define c_{ij} by $c_{ij} = -c_{ji}$ to form a skew symmetric matrix $C = (c_{ji})$ and put $\gamma = \sum_{i<j} c_{ji}\boldsymbol{e}_i \wedge \boldsymbol{e}_j$ in the form $\gamma = \frac{1}{2}\sum_{i,j} c_{ji}\boldsymbol{e}_i \wedge \boldsymbol{e}_j$. This implies that $\wedge^2\mathbb{R}^n$ can be identified with the space $\mathfrak{so}(n)$ of skew symmetric matrices by the isomorphism

$$R : \wedge^2\mathbb{R}^n \to \mathfrak{so}(n); \quad \gamma = \frac{1}{2}\sum_{i,j} c_{ji}\boldsymbol{e}_i \wedge \boldsymbol{e}_j \longmapsto C = (c_{ji}).$$

In particular, for $\boldsymbol{u} = \sum_i u_i\boldsymbol{e}_i$ and $\boldsymbol{v} = \sum_j v_j\boldsymbol{e}_j$, the wedge product $\boldsymbol{u} \wedge \boldsymbol{v}$ corresponds to

$$R(\boldsymbol{u} \wedge \boldsymbol{v}) = \frac{1}{2}\sum R((u_i v_j - u_j v_i)(\boldsymbol{e}_i \wedge \boldsymbol{e}_j)) = -\boldsymbol{u}\boldsymbol{v}^T + \boldsymbol{v}\boldsymbol{u}^T.$$

Since the inner product on $\mathfrak{so}(n)$ is defined through

$$\langle \xi, \eta \rangle = \frac{1}{2}\mathrm{tr}(\xi^T\eta), \quad \xi, \eta \in \mathfrak{so}(n),$$

the linear space $\wedge^2\mathbb{R}^n$ is endowed with the inner product so that $\wedge^2\mathbb{R}^n$ and $\mathfrak{so}(n)$ may be isometric to each other. In particular, the inner product of $\boldsymbol{u} \wedge \boldsymbol{v}$ and $\boldsymbol{x} \wedge \boldsymbol{y}$ is given by

$$\langle \boldsymbol{u} \wedge \boldsymbol{v}, \boldsymbol{x} \wedge \boldsymbol{y} \rangle = \langle R(\boldsymbol{u} \wedge \boldsymbol{v}), R(\boldsymbol{x} \wedge \boldsymbol{y}) \rangle = \det\begin{pmatrix} \boldsymbol{u} \cdot \boldsymbol{x} & \boldsymbol{u} \cdot \boldsymbol{y} \\ \boldsymbol{v} \cdot \boldsymbol{x} & \boldsymbol{v} \cdot \boldsymbol{y} \end{pmatrix},$$

where $\boldsymbol{x} \cdot \boldsymbol{y}$, etc., denote the inner product of $\boldsymbol{x}, \boldsymbol{y} \in \mathbb{R}^n$ and so on.

The $SO(n)$ action on $\mathbb{R}^n$ is naturally extended to that on $\wedge^2\mathbb{R}^n$,

$$\gamma = \sum_{i<j} c_{ji}\boldsymbol{e}_i \wedge \boldsymbol{e}_j \longmapsto g\gamma = \sum_{i<j} c_{ji} g\boldsymbol{e}_i \wedge g\boldsymbol{e}_j.$$

The following proposition is easy to prove.

Proposition 5.5.1 *The $SO(n)$ action on $\mathbb{R}^n$ and on $\wedge^2\mathbb{R}^n$, and the $\mathfrak{so}(n)$ action on $\mathbb{R}^n$ have the following properties:*

(1) $gR(\boldsymbol{x} \wedge \boldsymbol{y})g^{-1} = R(g\boldsymbol{x} \wedge g\boldsymbol{y}), \quad g \in SO(n),$
(2) $R(\theta)\boldsymbol{x} \cdot \boldsymbol{y} = \langle \theta, \boldsymbol{x} \wedge \boldsymbol{y} \rangle \quad \text{for} \quad \theta \in \wedge^2\mathbb{R}^n,$
(3) $R(\theta)\boldsymbol{x} \cdot R(\varphi)\boldsymbol{y} = \langle \theta, \boldsymbol{x} \wedge R(\varphi)\boldsymbol{y} \rangle \quad \text{for} \quad \theta, \varphi \in \wedge^2\mathbb{R}^n.$

Geometric setting on the center-of-mass system
The geometric setting for many-body systems in $\mathbb{R}^3$ can be extended in a natural manner to that for many-body systems in $\mathbb{R}^n$. Let $\boldsymbol{x}_1, \cdots, \boldsymbol{x}_N$ be position vectors of $\mathbb{R}^n$ and $m_1, \cdots, m_N$ positive numbers. The center-of-mass system is defined to be

$$Q = \{x = (\boldsymbol{x}_1, \cdots, \boldsymbol{x}_N) | \sum_{\alpha=1}^{N} m_\alpha \boldsymbol{x}_\alpha = 0\}.$$

The rotation group $SO(n)$ acts on Q in the manner

$$x = (\boldsymbol{x}_1, \cdots, \boldsymbol{x}_N) \mapsto gx = (g\boldsymbol{x}_1, \cdots, g\boldsymbol{x}_N), \quad g \in SO(n).$$

The property (1) of Proposition 5.5.1 is extended to be

$$gR(\gamma)g^{-1} = R(g\gamma), \quad \gamma \in \wedge^2\mathbb{R}^n, \quad g \in SO(n).$$

For $v = (\boldsymbol{v}_1, \cdots, \boldsymbol{v}_N) \in T_x(Q)$, the total angular momentum is defined to be

$$\sum m_\alpha \boldsymbol{x}_\alpha \wedge \boldsymbol{v}_\alpha.$$

Further, the mass-weighted metric (metric, for short) on $T_x(Q)$ is defined by

$$K_x(u, v) = \sum m_\alpha \boldsymbol{u}_\alpha \cdot \boldsymbol{v}_\alpha, \quad u, v \in T_x(Q).$$

The generalized inertia tensor $A_x : \wedge^2\mathbb{R}^n \to \wedge^2\mathbb{R}^n$ is defined to be

$$A_x(\theta) = \sum m_\alpha \boldsymbol{x}_\alpha \wedge R(\theta)\boldsymbol{x}_\alpha, \quad \theta \in \wedge^2\mathbb{R}^n,$$

where $x = (\boldsymbol{x}_1, \cdots, \boldsymbol{x}_N) \in Q$. This inertial tensor has similar properties to the inertia tensor for the many-body system in three dimensions.

Proposition 5.5.2 *If* $\dim \operatorname{span}\{\boldsymbol{x}_\alpha\} \geq n-1$, *the inertia tensor* A_x *is symmetric and positive-definite, and further has the properties shown below:*

(1) $\langle \theta, A_x(\varphi)\rangle = \langle A_x(\theta), \varphi\rangle, \quad \theta, \varphi \in \wedge^2\mathbb{R}^n,$
(2) $\langle \theta, A_x(\theta)\rangle \geq 0, \quad \langle \theta, A_x(\theta)\rangle = 0 \quad \Leftrightarrow \quad \theta = 0,$
(3) $R(A_x(\theta)) = J_x R(\theta) + R(\theta) J_x, \quad J_x := \sum_\alpha m_\alpha \boldsymbol{x}_\alpha \boldsymbol{x}_\alpha^T,$
(4) $A_{gx}(\theta) = g A_x(g^{-1}\theta), \quad \theta \in \wedge^2\mathbb{R}^n, \quad g \in SO(n).$

Proof As (1), (3), and (4) are easy to prove by straightforward calculation, we prove (2), the positive-definiteness of A_x under the condition $\dim \operatorname{span}\{\boldsymbol{x}_\alpha\} \geq n-1$. Since $\langle \theta, A_x(\theta)\rangle = \sum m_\alpha R(\theta)\boldsymbol{x}_\alpha \cdot R(\theta)\boldsymbol{x}_\alpha \geq 0$ on account of Proposition 5.5.1, we have only to show that

$$\langle \theta, A_x(\theta)\rangle = 0 \quad \Leftrightarrow \quad R(\theta) = 0,$$

where $R(\theta) = 0$ is equivalent to $\theta = 0$. We first note that $\langle \theta, A_x(\theta)\rangle = 0 \Leftrightarrow R(\theta)\boldsymbol{x}_\alpha = 0$, $\alpha = 1, \ldots, N$. Since $\dim \operatorname{span}\{\boldsymbol{x}_\alpha\} \geq n-1$, there exist $n-1$ linearly independent vectors $\boldsymbol{u}_k$, $k = 1, \ldots, n-1$, which span the subspace $\operatorname{span}\{\boldsymbol{x}_\alpha\}$. Then, the condition $R(\theta)\boldsymbol{x}_\alpha = 0$ leads to $R(\theta)\boldsymbol{u}_k = 0$. We can assume that $\boldsymbol{u}_k$, $k = 1, \ldots, n-1$, have been normalized. We add a normalized vector $\boldsymbol{u}_n$ in such a manner that $\boldsymbol{u}_k$, $k = 1, \ldots, n-1, n$, form an orthonormal basis of $\mathbb{R}^n$. Then, for $\boldsymbol{u}_n$, we obtain

$$R(\theta)\boldsymbol{u}_n \cdot \boldsymbol{u}_k = \boldsymbol{u}_n \cdot R(\theta)^T \boldsymbol{u}_k = -\boldsymbol{u}_n \cdot R(\theta)\boldsymbol{u}_k = 0, \quad k = 1, \ldots, n-1.$$

This implies that $R(\theta)\boldsymbol{u}_n$ is orthogonal to $\boldsymbol{u}_k$, $k = 1, \ldots, n-1$, so that $R(\theta)\boldsymbol{u}_n$ can be expressed as $R(\theta)\boldsymbol{u}_n = c\boldsymbol{u}_n$. Then, a further calculation provides

$$c = R(\theta)\boldsymbol{u}_n \cdot \boldsymbol{u}_n = -\boldsymbol{u}_n \cdot R(\theta)\boldsymbol{u}_n = -c,$$

which implies that $c = 0$. It then follows that $R(\theta)\boldsymbol{u}_k = 0$ for all $k = 1, \ldots, n$, which means that $R(\theta) = 0$. This ends the proof.

Connection and curvature
In what follows, we restrict ourselves to the subset of Q which is subject to the condition $\dim \operatorname{span}\{\boldsymbol{x}_\alpha\} \geq n-1$,

$$\dot{Q} = \left\{(x = (\boldsymbol{x}_1, \cdots, \boldsymbol{x}_N) | \sum m_\alpha \boldsymbol{x}_\alpha = 0, \ \dim \operatorname{span}\{\boldsymbol{x}_\alpha\} \geq n-1\right\}.$$

Then, the $SO(n)$ action on $\dot{Q}$ is shown to be free, i.e., if $gx = x$ for $x \in \dot{Q}$, then $g = I_n$, the identity of $SO(n)$. In fact, on taking an orthonormal basis, $\boldsymbol{u}, \ldots, \boldsymbol{u}_{n-1}, \boldsymbol{u}_n$, of $\mathbb{R}^n$ such that $\boldsymbol{u}, \ldots, \boldsymbol{u}_{n-1}$ span the subspace $\operatorname{span}\{\boldsymbol{x}_\alpha\}$ and $\boldsymbol{u}_n$ is orthogonal to it, the condition $gx = x$ means that $g\boldsymbol{u}_k = \boldsymbol{u}_k$, $k = 1, \ldots, n-1$. Since g is an orthogonal matrix, it follows that $g\boldsymbol{u}_n = \pm\boldsymbol{u}_n$. Since $\det(g) = 1$, one has $g\boldsymbol{u}_n = \boldsymbol{u}_n$, so that $g = I_n$. It then turns out that $\dot{Q}$ has a fiber bundle structure with the structure group $SO(n)$.

We proceed to the definition of rotational and vibrational vectors. For any $\theta \in \wedge^2\mathbb{R}^n$, a rotational vector at $x = (\boldsymbol{x}_1, \cdots, \boldsymbol{x}_N) \in \dot{Q}$ is defined to be

$$\frac{d}{dt}(e^{tR(\theta)}\boldsymbol{x}_1, \cdots, e^{tR(\theta)}\boldsymbol{x}_N)\Big|_{t=0} = (R(\theta)\boldsymbol{x}_1, \cdots, R(\theta)\boldsymbol{x}_N).$$

The rotational vector field corresponding to $R(\theta) \in \mathfrak{so}(n)$ is compactly denoted by

$$R(\theta)_{\dot{Q}}(x) = \frac{d}{dt}e^{tR(\theta)}x\Big|_{t=0}, \quad x \in \dot{Q},$$

which is also written as $R(\theta)x$. A tangent vector $v = (\boldsymbol{v}_1, \cdots, \boldsymbol{v}_N) \in T_x(\dot{Q})$ which is orthogonal to any rotational vector, i.e., $K_x(v, R(\theta)x) = 0$ for any $\theta \in \wedge^2\mathbb{R}^n$, is called a vibrational vector at x. It is easy to show that

$$v = (\boldsymbol{v}_1, \cdots, \boldsymbol{v}_N) \text{ is vibrational} \quad \Leftrightarrow \quad \sum m_\alpha \boldsymbol{x}_\alpha \wedge \boldsymbol{v}_\alpha = 0.$$

Thus, the tangent space is decomposed into the orthogonal direct sum of the rotational and the vibrational subspaces,

$$T_x(\dot{Q}) = W_{x,\mathrm{rot}} \oplus W_{x,\mathrm{vib}}.$$

As A_x is invertible for $x \in \dot{Q}$, the connection form ω on $\dot{Q}$ is now defined through

$$\omega_x(v) = R(A_x^{-1} \sum m_\alpha \boldsymbol{x}_\alpha \wedge \boldsymbol{v}_\alpha), \quad v = (\boldsymbol{v}_1, \cdots, \boldsymbol{v}_N) \in T_x(\dot{Q}).$$

It is straightforward to show the following:

Proposition 5.5.3 *The connection form on $\dot{Q}$ satisfies*

$$\omega_x(R(\theta)x) = R(\theta), \quad \omega_{gx}(gv) = \mathrm{Ad}_g \omega_x(v), \quad v \in T_x(\dot{Q}).$$

These equations are described also as

$$\omega(R(\theta)_{\dot{Q}}) = R(\theta), \quad \Phi_g^*\omega = \mathrm{Ad}_g\omega,$$

respectively, where Φ_g denotes the action of $SO(n)$ on $\dot{Q}$, $\Phi_g(x) = gx$, and $\Phi_g^*\omega$ the pull-back of ω by Φ_g.

The curvature form Ω is defined to be

$$\Omega = d\omega - \omega \wedge \omega,$$

which takes values for vector fields u and v,

$$\Omega(u, v) = d\omega(u, v) - [\omega(u), \omega(v)], \tag{5.29}$$

where $[\cdot, \cdot]$ denotes the bracket operator defined on $\mathfrak{so}(n)$.

Proposition 5.5.4 *The curvature form has the properties:*

(i) $\Phi_g^*\Omega = \mathrm{Ad}_g\Omega$,
(ii) $\Omega_x(u, v) = 0 \quad \Leftrightarrow \quad u \text{ or } v \in W_{x,\mathrm{rot}}$.

Proof The property (i) is a consequence of the similar property, $\Phi_g^*\omega = \mathrm{Ad}_g\omega$, of the connection form. We prove the property (ii). For $R(\theta) \in \mathfrak{so}(n)$ with $\theta \in \wedge^2\mathbb{R}^n$ and for $g_t = e^{tR(\theta)} \in SO(n)$, the equality $\Phi_{g_t}^*\omega = \mathrm{Ad}_{g_t}\omega$ is differentiated with respect to t at $t = 0$ to provide

$$\mathcal{L}_{R(\theta)_{\dot{Q}}}\omega = [R(\theta), \omega],$$

where the left-hand side is the Lie derivative of ω with respect to the vector field $R(\theta)_{\dot{Q}}$ [1]. On the other hand, by the use of the Cartan formula about the Lie derivative [1], the Lie derivative $\mathcal{L}_{R(\theta)_{\dot{Q}}}\omega$ is arranged as

$$\mathcal{L}_{R(\theta)_{\dot{Q}}}\omega = d\iota(R(\theta)_{\dot{Q}})\omega + \iota(R(\theta)_{\dot{Q}})d\omega = \iota(R(\theta)_{\dot{Q}})d\omega,$$

where use has been made of $\omega(R(\theta)_{\dot{Q}}) = R(\theta)$. Hence, we obtain

$$\iota(R(\theta)_{\dot{Q}})d\omega = [R(\theta), \omega] = [\omega(R(\theta)_{\dot{Q}}), \omega]. \tag{5.30}$$

From (5.29) and (5.30), it follows that

$$\Omega(R(\theta)_{\dot{Q}}, v) = d\omega(R(\theta)_{\dot{Q}}, v) - [\omega(R(\theta)_{\dot{Q}}), \omega(v)] = 0.$$

This ends the proof.

In the rest of this section, we show that the curvature Ω never vanishes on $\dot{Q}$. Applying the formula $d\omega(X, Y) = X(\omega(Y)) - Y(\omega(X)) - \omega([X, Y])$ for vibrational vector fields u, v, one has

$$\Omega(u, v) = d\omega(u, v) = -\omega([u, v]).$$

Thus, what we have to do is to show that the right-hand side of the above equation does not vanish for vibrational vector fields. We start by calculating the bracket $[u, v]$. For notational convenience, we denote the components of x by

$$x = (\boldsymbol{x}_1, \cdots, \boldsymbol{x}_N) = (x_\alpha^i), \quad 1 \le i \le n, \quad 1 \le \alpha \le N,$$

and vibrational vectors $u = (\boldsymbol{u}_1, \cdots, \boldsymbol{u}_N)$ and $v = (\boldsymbol{v}_1, \cdots, \boldsymbol{v}_N)$ by

$$u = \sum_{i,\alpha} u^i_\alpha \frac{\partial}{\partial x^i_\alpha}, \qquad v = \sum_{i,a} v^i_\alpha \frac{\partial}{\partial x^i_\alpha},$$

respectively. Since u and v are vibrational, they should be subject to the conditions

$$\sum_\alpha m_\alpha \boldsymbol{x}_\alpha \wedge \boldsymbol{u}_\alpha = \sum_{i<j} \sum_\alpha m_\alpha (x^i_\alpha u^j_\alpha - x^j_\alpha u^i_\alpha) \boldsymbol{e}_i \wedge \boldsymbol{e}_j = 0,$$

$$\sum_\alpha m_\alpha \boldsymbol{x}_\alpha \wedge \boldsymbol{v}_\alpha = \sum_{i<j} \sum_\alpha m_\alpha (x^i_\alpha v^j_\alpha - x^j_\alpha v^i_\alpha) \boldsymbol{e}_i \wedge \boldsymbol{e}_j = 0,$$

respectively. Taking the derivative of the upper (resp. lower) equation with respect to v (resp. u) and putting the resulting equations together by reference to the definition of $[u, v]$,

$$[u, v] = \sum [u, v]^i_\alpha \frac{\partial}{\partial x^i_\alpha}, \qquad [u, v]^i_\alpha := \sum_{k,\beta} \left(u^k_\beta \frac{\partial v^i_\alpha}{\partial x^k_\beta} - v^k_\beta \frac{\partial u^i_\alpha}{\partial x^k_\beta} \right),$$

we find that

$$\sum_\alpha m_\alpha \boldsymbol{x}_\alpha \wedge [u, v]_\alpha = -2 \sum_\alpha m_\alpha \boldsymbol{u}_\alpha \wedge \boldsymbol{v}_\alpha.$$

By using this equation, we evaluate $\Omega(u, v)$ to obtain

$$\begin{aligned} \Omega(u, v) &= -\omega([u, v]) = -R(A_x^{-1} \sum m_\alpha \boldsymbol{x}_\alpha \wedge [u, v]_\alpha) \\ &= 2R(A_x^{-1} \sum m_\alpha \boldsymbol{u}_\alpha \wedge \boldsymbol{v}_\alpha), \end{aligned}$$

which suggests that our last task is to know what subspace is spanned by $\sum m_\alpha \boldsymbol{u}_\alpha \wedge \boldsymbol{v}_\alpha$ for all vibrational vectors u and v. According to [22], we know the following.

Proposition 5.5.5 *Let $x = (\boldsymbol{x}_1, \cdots, \boldsymbol{x}_N) \in \dot{Q}$ be a configuration such that* $\operatorname{span}\{\boldsymbol{x}_\alpha\} = \mathbb{R}^n$ *with $N > n \geq 3$. When tangent vectors v and v' run over the vibrational subspace $W_{x,\mathrm{vib}}$, the two-vectors $\sum m_\alpha \boldsymbol{v}_\alpha \wedge \boldsymbol{v}'_\alpha$ generate $\wedge^2\mathbb{R}^n$.*

Proof Since $\operatorname{span}\{\boldsymbol{x}_1, \cdots, \boldsymbol{x}_N\} = \mathbb{R}^n$, we may assume that $\boldsymbol{x}_1, \cdots, \boldsymbol{x}_n$ are a basis of $\mathbb{R}^n$, by renumbering the position vectors, if necessary. Then, $\boldsymbol{x}_\alpha \wedge \boldsymbol{x}_\beta$ with $1 \leq \alpha < \beta \leq n$ form a basis of $\wedge^2\mathbb{R}^n$. For our purpose, it is sufficient to show that $\boldsymbol{x}_1 \wedge \boldsymbol{x}_2$ can be represented in the form of $\sum m_\alpha \boldsymbol{v}_\alpha \wedge \boldsymbol{v}'_\alpha$ with $v, v' \in W_{x,\mathrm{vib}}$. If we want to show that $\boldsymbol{x}_2 \wedge \boldsymbol{x}_4$ can be also represented in a similar form, we have only to replace $\boldsymbol{x}_3$ for $\boldsymbol{x}_1$ and $\boldsymbol{x}_4$ for $\boldsymbol{x}_2$ in the course of the reasoning. Let

$$\boldsymbol{x}_{n+1} = \sum_{\alpha=1}^{n} r_\alpha \boldsymbol{x}_\alpha, \quad r = \sum_{\alpha=3}^{n} r_\alpha - 1.$$

Now, we introduce vectors $\boldsymbol{v}_1, \cdots, \boldsymbol{v}_N \in \mathbb{R}^n$ depending on arbitrary real constants, a_1, a_2, a_3, through

$$\begin{aligned}
m_1\boldsymbol{v}_1 &= \Big(\frac{1}{2}(-a_3 + r_2a_1 - r_1a_2) + ra_1\Big)\boldsymbol{x}_1 + \frac{1}{2}(a_3 - r_2a_1 + r_1a_2)\boldsymbol{x}_2,\\
m_2\boldsymbol{v}_2 &= \frac{1}{2}(a_3 - r_2a_1 + r_1a_2)\boldsymbol{x}_1 + \Big(\frac{1}{2}(-a_3 + r_2a_1 - r_1a_2) + ra_2\Big)\boldsymbol{x}_2,\\
m_\beta\boldsymbol{v}_\beta &= -r_\beta(a_1\boldsymbol{x}_1 + a_2\boldsymbol{x}_2), \quad \beta = 3, \ldots, n,\\
m_{n+1}\boldsymbol{v}_{n+1} &= a_1\boldsymbol{x}_1 + a_2\boldsymbol{x}_2,\\
m_\beta\boldsymbol{v}_\beta &= 0, \quad \beta > n+1.
\end{aligned}$$

For these vectors, one easily verifies that $v = (\boldsymbol{v}_1, \cdots, \boldsymbol{v}_N) \in W_{x,\mathrm{vib}}$, i.e.,

$$\sum_a m_\alpha \boldsymbol{v}_\alpha = 0, \quad \sum_\alpha m_\alpha \boldsymbol{x}_\alpha \wedge \boldsymbol{v}_\alpha = 0.$$

In the same manner, we can define $v' = (\boldsymbol{v}'_1, \cdots, \boldsymbol{v}'_N) \in W_{x,\mathrm{vib}}$ with a_1, a_2, a_3 replaced by a'_1, a'_2, a'_3, respectively. After a straightforward calculation, we verify that

$$\sum_\alpha m_\alpha \boldsymbol{v}_\alpha \wedge \boldsymbol{v}'_\alpha = \sum_{1\le i<j\le 3} c_{ij}(a_ia'_j - a_ja'_i)\boldsymbol{x}_1 \wedge \boldsymbol{x}_2,$$

where

$$c_{12} = \frac{r}{2}\Big(\frac{r_1}{m_1} - \frac{r_2}{m_2}\Big) + \sum_{\alpha=3}^{n} \frac{r_\alpha^2}{m_\alpha} + \frac{1}{m_{n+1}}, \quad c_{13} = \frac{r}{2m_1}, \quad c_{23} = -\frac{r}{2m_2}.$$

The quantities c_{12}, c_{13}, c_{23} do not vanish simultaneously. In fact, if $c_{13} = c_{23} = 0$, then one has

$$c_{12} = \sum_{\alpha=3}^{n} \frac{r_a^2}{m_a} + \frac{1}{m_{n+1}} > 0.$$

Thus, the factor $\sum c_{ij}(a_ia'_j - a_ja'_i)$ does not vanish, so that $\boldsymbol{x}_1 \wedge \boldsymbol{x}_2$ is expressed as a constant multiple of $\sum m_\alpha \boldsymbol{v}_\alpha \wedge \boldsymbol{v}'_\alpha$. This ends the proof.

Since $\omega([u, v])$ takes values in $\mathfrak{so}(n)$, it induces a vector field on $\dot{Q}$, which we denote by $\omega([u, v])_{\dot{Q}}$. For vibrational vectors u and v, the vector field $\omega([u, v])_{\dot{Q}}$ is considered as the rotational component of $[u, v]$. Since $\omega([u, v])$ does not necessarily vanish on account of Proposition 5.5.5, it follows that infinitesimal vibrational motions are coupled together to bring about an infinitesimal rotation.

Remark It is worth pointing out that the planar many-body systems can be revisited from the point of view taken in this section. Let $n = 2$. Then, one has $\dim \wedge^2\mathbb{R}^2 = 1$, and the canonical basis of $\wedge^2\mathbb{R}^2$ is $\boldsymbol{e}_1 \wedge \boldsymbol{e}_2$. The map $R : \wedge^2\mathbb{R}^2 \to \mathfrak{so}(2)$ is simply determined by

$$R(\boldsymbol{e}_1 \wedge \boldsymbol{e}_2) = \begin{pmatrix} 0 & -1 \\ 1 & 0 \end{pmatrix}.$$

For the planar three-body system, the Jacobi vectors are given by $\boldsymbol{r}_1 = q_1\boldsymbol{e}_1 + q_2\boldsymbol{e}_2$ and $\boldsymbol{r}_2 = q_3\boldsymbol{e}_1 + q_4\boldsymbol{e}_2$, and the inertia tensor is defined and evaluated as

$$A_x(\boldsymbol{e}_1 \wedge \boldsymbol{e}_2) = \sum_{j=1}^{2} \boldsymbol{r}_j \wedge R(\boldsymbol{e}_1 \wedge \boldsymbol{e}_2)\boldsymbol{r}_j = r^2 R(\boldsymbol{e}_1 \wedge \boldsymbol{e}_2), \quad r^2 = \sum_{k=1}^{4} q_k^2.$$

The connection form is defined and evaluated as

$$\omega_x = R\Big(A_x^{-1}\Big(\sum_{j=1}^{2} \boldsymbol{r}_j \wedge d\boldsymbol{r}_j\Big)\Big) = \frac{1}{r^2}(-q_2dq_1 + q_1dq_2 - q_4dq_3 + q_3dq_4)R(\boldsymbol{e}_1 \wedge \boldsymbol{e}_2).$$

Thus, we have obtained the connection form ω given in (1.61) again. Let σ be a local section. Then, for $\omega_{g\sigma(q)}$, we can find a formula similar to (1.187)

$$\omega_{g\sigma(q)} = g(g^{-1}dg + \omega_{\sigma(q)})g^{-1}. \tag{5.31}$$

For $g = \exp(\theta R(\boldsymbol{e}_1 \wedge \boldsymbol{e}_2))$, one obtains $g^{-1}R(\boldsymbol{e}_1 \wedge \boldsymbol{e}_2)g^{-1} = R(\boldsymbol{e}_1 \wedge \boldsymbol{e}_2)$ and $g^{-1}dg = d\theta R(\boldsymbol{e}_1 \wedge \boldsymbol{e}_2)$, which simplify the expression of the above equation.

5.6 Holonomy for Many-Body Systems

Let $c(t)$, $0 \le t \le 1$, be a (piecewise) smooth curve in the shape space $\dot{M} = \dot{Q}/SO(n)$. A vibrational (or horizontal) curve $c^*(t)$, $0 \le t \le 1$, in $\dot{Q}$ satisfying

$$\pi(c^*(t)) = c(t)$$

is called a vibrational (or horizontal) lift of $c(t)$. If a point $x_0 \in \pi^{-1}(c_0)$ with $c_0 = c(0)$ is given, there exists a unique horizontal lift $c^*(t)$, $0 \le t \le 1$, of $c(t)$ with

$c^*(0) = x_0$. In fact, for an arbitrary smooth curve $x(t) \in \dot{Q}$ such that $\pi(x(t)) = c(t)$ and $x_0 = x(0)$, we can determine a vibrational (or horizontal) curve $y(t)$ satisfying $x(t) = g(t)y(t)$ by solving the differential equation

$$\frac{dg}{dt} = \omega_x(\dot{x})g, \quad g(0) = \mathrm{id}_{SO(n)}.$$

We note that this equation is of the same form as Eq. (1.228) studied for many-body systems in three dimensions.

Let $c(t),\ 0 \leq t \leq 1$, be a loop, i.e., a closed curve. Then, there exists a vibrational (or horizontal) lift $c^*(t)$. For an arbitrary $x_0 \in \pi^{-1}(c_0)$, there exists a $h \in SO(n)$ such that $c^*(1) = hc^*(0)$ and $c^*(0) = x_0$. For all loops starting (and ending) with a fixed reference point $c_0 \in \dot{M}$ and a fixed initial point $x_0 \in \pi^{-1}(c_0)$, the set of h such that $c^*(1) = hc^*(0)$, $x_0 = c^*(0)$ form a subgroup of $SO(n)$, which is denoted by $H(x_0)$ and called a holonomy group with the reference point x_0. It is known that if $\dot{M}$ is connected then $H(x_0)$ is a Lie group. In fact, $\dot{M} = \dot{Q}/SO(n)$ is connected in the case of many-body systems. In addition, if $\dot{M}$ is connected, every holonomy group $H(x)$, $x \in \dot{Q}$ is conjugate, and then isomorphic.

Theorem 5.6.1 (Holonomy theorem [48]) *Let $\pi : P \to M$ be a principal G-bundle and M be connected. Let Ω denote the curvature form of the connection on P. Let $H(x)$ denote the holonomy group with a reference point x. Then, the Lie algebra $\mathfrak{H}(x)$ of $H(x)$ forms the subspace of the Lie algebra $\mathfrak{G}$ of G which is generated by all the elements of the form $\Omega_y(u, v)$, where y is a point which can be jointed to x by a horizontal curve and u and v are horizontal vectors at y.*

On the basis of the holonomy theorem, we can prove the following theorem [22]:

Theorem 5.6.2 *Let x and g be an arbitrary point of the center-of-mass system $\dot{Q}$ and an arbitrary point of $SO(n)$, respectively. Then, there exist a vibrational curve which joins x and gx, where $N > n$.*

Proof Let $c(t), 0 \leq t \leq 1$, be a loop in the shape space $\dot{M}$. For $c_0 = c(0)$, we take an arbitrary point $x_0 \in \pi^{-1}(c_0)$. Let $H(x_0)$ denote the holonomy group with the reference point x_0. The Lie algebra $\mathfrak{H}(x_0)$ of $H(x_0)$ contains all the elements of the form $\Omega_{x_0}(v, v')$ with $v, v' \in W_{x_0,\mathrm{vib}}$, which forms a subalgebra of $\mathfrak{so}(n)$.

We may take x_0 as consisting of position vectors such that $\boldsymbol{x}_1, \cdots, \boldsymbol{x}_N$ span $\mathbb{R}^n$. Then, Proposition 5.5.5 implies that $\sum m_\alpha \boldsymbol{v}_\alpha \wedge \boldsymbol{v}'_\alpha$ with $v, v' \in W_{x_0,\mathrm{vib}}$ generate $\wedge^2\mathbb{R}^n \cong \mathfrak{so}(n)$. Hence, when v and v' run over $W_{x_0,\mathrm{vib}}$, $\Omega_{x_0}(v, v') = 2R(A_{x_0}^{-1} \sum m_\alpha \boldsymbol{v}_\alpha \wedge \boldsymbol{v}'_\alpha)$ generate $\mathfrak{so}(n)$, so that $\mathfrak{H}(x_0) = \mathfrak{so}(n)$. This implies that $H(x_0) = SO(n)$. Since $\dot{M}$ is connected, all the holonomy groups are conjugate, so that for every $x \in \dot{Q}$, one has $H(x) = SO(n)$. This completes the proof.

As a corollary, if $N > n$, any points x and y in the center-of-mass system can be jointed by a vibrational curve.

5.7 Rigid Bodies in n Dimensions

The inertia tensor
Mechanics for a free rigid body in $\mathbb{R}^n$ can be naturally set up, in a similar manner to that for a rigid body in $\mathbb{R}^3$, on the basis of the N-body system dealt with in Sec. 5.5. If the N-body system in $\mathbb{R}^n$ is supposed to be frozen, it is viewed as a rigid body. Then, from (3) of Proposition 5.5.2, the inertia tensor of the rigid body is defined to be the map $\mathcal{A} : \mathfrak{so}(n) \to \mathfrak{so}(n)$ given by

$$\mathcal{A}(\xi) = B\xi + \xi B, \quad \xi \in \mathfrak{so}(n),$$

where B is a symmetric positive-definite $n \times n$ matrix, which comes from $J_x = \sum_\alpha m_\alpha \boldsymbol{x}_\alpha \boldsymbol{x}_\alpha^T$. Like (1.86), we define the inner product on $\mathfrak{so}(n)$ to be

$$\langle \xi, \eta \rangle = \frac{1}{2}\mathrm{tr}(\xi^T \eta), \quad \xi, \eta \in \mathfrak{so}(n).$$

Then, it is easily verified that

$$\langle [\xi, \eta], \zeta \rangle = \langle \xi, [\eta, \zeta] \rangle, \quad \xi, \eta, \zeta \in \mathfrak{so}(n) \tag{5.32}$$

and

$$\langle \mathrm{Ad}_h(\xi), \mathrm{Ad}_h(\eta) \rangle = \langle \xi, \eta \rangle, \quad \xi, \eta \in \mathfrak{so}(n), \quad h \in SO(n).$$

With respect to the present inner product, the inertia tensor $\mathcal{A}$ is symmetric and positive-definite. We show the positive-definiteness only in what follows. From the definition of $\mathcal{A}$, one obtains $\langle \xi, \mathcal{A}(\xi) \rangle = \mathrm{tr}(B\xi\xi^T)$. Since B is positive-definite, there exists an orthogonal matrix h such that $B = hDh^{-1}$, $D = \mathrm{diag}(\lambda_1, \cdots, \lambda_n)$, $\lambda_j > 0$. Then, the $\mathrm{tr}(B\xi\xi^T)$ becomes

$$\mathrm{tr}(hDh^{-1}\xi\xi^T) = \mathrm{tr}(Dh^{-1}\xi(h^{-1}\xi)^T) = \mathrm{tr}(D\zeta\zeta^T) = \sum_{i,k} \lambda_i \zeta_{ik}^2 \geq 0,$$

where $\zeta = h^{-1}\xi$, $\zeta = (\zeta_{ij})$. The above equation means that $\mathcal{A}$ is positive-semi-definite. Now suppose that $\mathrm{tr}(B\xi\xi^T) = 0$. Then, the above equation means that $\sum_{i,k} \lambda_i \zeta_{ik}^2 = 0$. Since $\lambda_i > 0$, one has $\zeta_{ik} = 0$, so that $\zeta = h^{-1}\xi = 0$ and hence $\xi = 0$. Thus, $\mathcal{A}$ proves to be positive-definite.

The eigenvalues of $\mathcal{A}$ and those of B are related in a simple manner, as is shown below. Let $\xi \in \mathfrak{so}(n)$ be an eigenvector (matrix) associated with the eigenvalue μ of $\mathcal{A}$. Then one has

$$\mathcal{A}(\xi) = B\xi + \xi B = \mu\xi.$$

Since B is a symmetric matrix, it is written out as $B = hDh^{-1}$ with $D = \mathrm{diag}(\lambda_1, \cdots, \lambda_n)$ for $h \in O(n)$. Then, the above eigenvalue equation becomes $hDh^{-1}\xi + \xi hDh^{-1} = \mu\xi$. Multiplying this equation by h^{-1} and h to the left and the right, respectively, results in

$$Dh^{-1}\xi h + h^{-1}\xi h D = \mu h^{-1}\xi h.$$

Setting $h^{-1}\xi h = \eta = (\eta_{ij})$, we rewrite this equation as $D\eta + \eta D = \mu\eta$, which is put, componentwise, in the form

$$\sum_j \lambda_i \delta_{ij}\eta_{jk} + \sum_j \eta_{ij}\lambda_j\delta_{jk} = \mu\eta_{ik},$$

so that

$$(\lambda_i + \lambda_j)\eta_{ik} = \mu\eta_{ik}.$$

If $\eta_{ik} \neq 0$, then $\lambda_i + \lambda_j = \mu$. Since η is anti-symmetric, we may assume that $i < j$. Let E_{ij} be a matrix whose (i, j) component is 1 and the others are 0. Then, for $\eta = E_{ij} - E_{ji}$, the matrix $\xi = h\eta h^{-1} = h(E_{ij} - E_{ji})h^{-1}$ gives an eigenvector (matrix) associated with the eigenvalue $\lambda_i + \lambda_j$. We now recall that $h\boldsymbol{e}_i$ is an eigenvector of B. Since $E_{ij} = \boldsymbol{e}_i\boldsymbol{e}_i^T$, the eigenvector (matrix) in question is expressed also as

$$h(E_{ij} - E_{ji})h^{-1} = h(\boldsymbol{e}_i\boldsymbol{e}_j^T - h\boldsymbol{e}_j\boldsymbol{e}_i^T)h^T = h\boldsymbol{e}_i(h\boldsymbol{e}_j)^T - h\boldsymbol{e}_j(h\boldsymbol{e}_i)^T, \tag{5.33}$$

which shows that the eigenvectors (matrices) of $\mathcal{A}$ are described in terms of the eigenvectors $h\boldsymbol{e}_i$ of B.

Equations of motion

Now we are in a position to describe mechanics for rigid bodies in n dimensions. Like (2.17), the Lagrangian is defined on $SO(n) \times \mathfrak{so}(n)$ to be

$$L(g, \xi) = \frac{1}{2}\langle \xi, \mathcal{A}(\xi)\rangle, \quad \xi = g^{-1}\dot{g} \in \mathfrak{so}(n).$$

The derivation of Lagrange's equation of motion runs in parallel to that for the rigid body in $\mathbb{R}^3$. On using $\delta\xi = [\xi, g^{-1}\delta g] + \dfrac{d}{dt}(g^{-1}\delta g)$, the infinitesimal variation of L is written out as

$$\begin{aligned}\delta L &= \frac{1}{2}\langle \delta\xi, \mathcal{A}(\xi)\rangle + \frac{1}{2}\langle \xi, \mathcal{A}(\delta\xi)\rangle = \langle \delta\xi, \mathcal{A}(\xi)\rangle \\ &= \left\langle \frac{d}{dt}(g^{-1}\delta g), \mathcal{A}(\xi)\right\rangle + \left\langle [\xi, g^{-1}\delta g], \mathcal{A}(\xi)\right\rangle \\ &= \langle g^{-1}\delta g, -\frac{d}{dt}\mathcal{A}(\xi)\rangle + \langle [\mathcal{A}(\xi), \xi], g^{-1}\delta g\rangle + \frac{d}{dt}\langle g^{-1}\delta g, \mathcal{A}(\xi)\rangle,\end{aligned}$$

where (5.32) has been used. Since $g^{-1}\delta g$ is arbitrary and since $\delta g(t_1) = \delta g(t_2) = 0$, it follows that

$$\delta \int_{t_1}^{t_2} L(g, \xi)dt = 0 \quad \Leftrightarrow \quad \frac{d}{dt}\mathcal{A}(\xi) = [\mathcal{A}(\xi), \xi], \tag{5.34}$$

where the right-hand side of the above equation, the n-dimensional Euler equation, is supplemented with $dg/dt = g\xi$. This equation is clearly a generalization of Eq. (2.21).

Like the Euler equation for a rigid body in $\mathbb{R}^3$, the equation of motion is equivalent to the conservation of the angular momentum $g\mathcal{A}(\xi)g^{-1}$. In fact, under the condition $\xi = g^{-1}\dot{g}$, one verifies that

$$\frac{d}{dt}(g\mathcal{A}(\xi)g^{-1}) = 0 \quad \Leftrightarrow \quad \frac{d}{dt}\mathcal{A}(\xi) = [\mathcal{A}(\xi), \xi].$$

We proceed to Hamiltonian mechanics for the rigid body in n dimensions. Like (2.35), the functional derivative, $\dfrac{\delta F}{\delta \xi}(\xi)$, of F on $\mathfrak{so}(n)$ is defined through

$$\left\langle \frac{\delta F}{\delta \xi}(\xi), \eta \right\rangle = \frac{d}{dt}F(\xi + t\eta)|_{t=0}, \quad \eta \in \mathfrak{so}(n).$$

Introducing $\zeta = \frac{\delta L}{\delta \xi}$ as a conjugate variable to ξ, we define the Hamiltonian to be $H(g, \zeta) = \langle \zeta, \xi \rangle - L(g, \xi)$. Since $\zeta = \mathcal{A}(\xi)$, the Hamiltonian on $SO(n) \times \mathfrak{so}(n)$ is expressed as

$$H(g, \zeta) = \frac{1}{2}\langle \zeta, \mathcal{A}^{-1}(\zeta)\rangle, \quad (g, \zeta) \in SO(3) \times \mathfrak{so}(n),$$

where the dual space $\mathfrak{so}(n)^*$ is identified with $\mathfrak{so}(n)$ through the inner product on $\mathfrak{so}(n)$.

The principle of variation applied to

$$\int_{t_1}^{t_2} (\langle \zeta, \xi \rangle - H)dt, \quad \xi = g^{-1}\frac{dg}{dt}$$

will bring about Hamilton's equations of motion under the boundary conditions $\delta g(t_1) = \delta g(t_2) = 0$ and $\delta \zeta(t_1) = \delta \zeta(t_2) = 0$. The infinitesimal variation of the above integral is evaluated as

$$\delta \int_{t_1}^{t_2} (\langle \zeta, \xi \rangle - H)dt = \int_{t_1}^{t_2} (\langle \delta\zeta, \xi \rangle + \langle \zeta, \delta\xi \rangle - \delta H)dt$$

$$= \int_{t_1}^{t_2} \left(\langle \delta\zeta, \xi \rangle + \left\langle \zeta, [\xi, g^{-1}\delta g] + \frac{d}{dt}(g^{-1}\delta g) \right\rangle - \left\langle \frac{\delta H}{\delta \zeta}, \delta\zeta \right\rangle \right) dt$$

$$= \int_{t_1}^{t_2} \left(\left\langle \xi - \frac{\delta H}{\delta \zeta}, \delta\zeta \right\rangle + \left\langle [\zeta, \xi] - \frac{d\zeta}{dt}, g^{-1}\delta g \right\rangle \right) dt.$$

Since $\delta\zeta$ and $g^{-1}\delta g$ are arbitrary, the principle of variation provides

$$\xi = \frac{\delta H}{\delta \zeta}, \quad [\zeta, \xi] - \frac{d\zeta}{dt} = 0.$$

Since $\delta H/\delta\zeta = \mathcal{A}^{-1}(\zeta)$, one obtains Hamilton's equations of motion

$$\frac{dg}{dt} = g\mathcal{A}^{-1}(\zeta), \quad \frac{d\zeta}{dt} = [\zeta, \mathcal{A}^{-1}(\zeta)], \tag{5.35}$$

which are the Euler equations for a rigid body in $\mathbb{R}^n$. It is clear that Eq. (5.35) is a generalization of Eq. (2.28).

Relative equilibria

Though it is difficult in general to solve Hamilton's equations of motion, there are solutions of special type, which are easy to find. Let ζ_0 be an eigenvector (matrix) of $\mathcal{A}^{-1}$. Then, one has $[\zeta_0, \mathcal{A}^{-1}(\zeta_0)] = 0$, so that $\zeta(t) = \zeta_0$ is a solution to the second of the above equations. The first of the above equations is now rewritten as

$$\frac{dg}{dt} = g\mathcal{A}^{-1}(\zeta_0),$$

which is easy to integrate. If ζ_0 is an eigenvector (matrix) of $\mathcal{A}^{-1}$, it is also an eigenvector (matrix) of $\mathcal{A}$. We now recall that there exists an eigenvector such that

$$\mathcal{A}(\zeta_0) = (\lambda_i + \lambda_j)\zeta_0, \quad \zeta_0 = h(E_{ij} - E_{ji})h^{-1}, \quad h \in O(n).$$

This leads to $\mathcal{A}^{-1}(\zeta_0) = \zeta_0/(\lambda_i + \lambda_j)$, so that the differential equation for g is expressed as

$$\frac{dg}{dt} = \frac{1}{\lambda_i + \lambda_j} g\zeta_0,$$

which is easily integrated to provide a solution $g(t) = g(0)\exp\left(\frac{t}{\lambda_i+\lambda_j}\zeta_0\right)$. In particular, from $\zeta_0 = h(E_{ij} - E_{ji})h^{-1}$, one has

$$\exp\left(\frac{t}{\lambda_i + \lambda_j}\zeta_0\right) = h\exp\left(\frac{t}{\lambda_i + \lambda_j}(E_{ij} - E_{ji})\right)h^{-1},$$

where

$$\exp\left(\frac{t}{\lambda_i + \lambda_j}(E_{ij} - E_{ji})\right) = \begin{pmatrix} \ddots & & & & \\ & \cos\frac{t}{\lambda_i+\lambda_j} & \cdots & \sin\frac{t}{\lambda_i+\lambda_j} & \\ & \vdots & \ddots & \vdots & \\ & -\sin\frac{t}{\lambda_i+\lambda_j} & \cdots & \cos\frac{t}{\lambda_i+\lambda_j} & \\ & & & & \ddots \end{pmatrix}, \quad i < j.$$

Hence, the period T of $g(t)$ is given by $T = 2\pi(\lambda_i + \lambda_j)$. The solution $(g(t), \zeta_0)$ in $SO(n) \times \mathfrak{so}(n)$ is called a relative equilibrium [25].

In the rest of this section, we are interested in the stability of the equilibrium $\zeta(t) = \zeta_0$. As is easily verified, there exist constants of motion,

$$\mathcal{L} = g\zeta g^{-1}, \quad H = \frac{1}{2}\langle \zeta, \mathcal{A}^{-1}(\zeta)\rangle, \quad M = \langle \zeta, \zeta\rangle,$$

where we note that $M = \langle \mathcal{L}, \mathcal{L}\rangle$.

Since M is a constant of motion, the second equation of (5.35) can be viewed as a differential equation on the sphere S^k determined by $M = 1$ in $\mathfrak{so}(n) = \mathbb{R}^{\frac{1}{2}n(n-1)}$, where $k = \frac{1}{2}n(n-1) - 1$, and further, H is considered as a function on the sphere S^k. The gradient, ∇H, of H on this sphere is defined through

$$dH = \langle \nabla H, d\zeta\rangle, \quad \langle \zeta, d\zeta\rangle = 0.$$

From the definition of H together with the constraint $\langle \zeta, \zeta\rangle = 1$, dH is evaluated as

$$dH = \langle d\zeta, \mathcal{A}^{-1}(\zeta)\rangle, \quad \langle d\zeta, \zeta\rangle = 0,$$

so that ∇H is put in the form $\nabla H = \mathcal{A}^{-1}(\zeta) + \lambda\zeta$, where λ is a Lagrange multiplier. Since ∇H is tangent to the sphere S^k, it should satisfy

$$\langle \nabla H, \zeta\rangle = \langle \mathcal{A}^{-1}(\zeta), \zeta\rangle + \lambda\langle \zeta, \zeta\rangle = 0,$$

which determines λ. It then follows that

$$\nabla H(\zeta) = \mathcal{A}^{-1}(\zeta) - \langle \mathcal{A}^{-1}(\zeta), \zeta\rangle\zeta, \quad \langle \zeta, \zeta\rangle = 1.$$

Let $\zeta_0 \in \mathfrak{so}(n)$ be an eigenvector associated with an eigenvalue μ_0 of $\mathcal{A}^{-1}$, where $\langle \zeta_0, \zeta_0\rangle = 1$. Let η be a tangent vector to S^k at ζ_0, that is, $\eta \in \mathfrak{so}(n)$, $\langle \zeta_0, \eta\rangle = 0$. In order to obtain the directional derivative, $\nabla^2 H(\eta)$, of ∇H in the direction of η in the form of a linear transformation on the tangent space $T_{\zeta_0}(S^k)$, we set $\zeta = \zeta_0 + \varepsilon\eta$ with ε an infinitesimal parameter, and evaluate $\nabla H(\zeta_0 + \varepsilon\eta)$ to the first order of ε to obtain

$$\begin{aligned}
&\nabla H(\zeta_0 + \varepsilon\eta) = \mathcal{A}^{-1}(\zeta_0 + \varepsilon\eta) - \langle \mathcal{A}^{-1}(\zeta_0 + \varepsilon\eta), \zeta_0 + \varepsilon\eta\rangle(\zeta_0 + \varepsilon\eta) \\
&= \mathcal{A}^{-1}(\zeta_0) + \varepsilon\mathcal{A}^{-1}(\eta) \\
&\quad -\big(\langle \mathcal{A}^{-1}(\zeta_0), \zeta_0\rangle + \varepsilon\langle \mathcal{A}^{-1}(\zeta_0), \eta\rangle + \varepsilon\langle \mathcal{A}^{-1}(\eta), \zeta_0\rangle\big)(\zeta_0 + \varepsilon\eta) \\
&= \varepsilon\big(\mathcal{A}^{-1}(\eta) - \mu_0\eta\big),
\end{aligned}$$

where use has been made of $\langle \zeta_0, \zeta_0 \rangle = 1$, $\langle \eta, \zeta_0 \rangle = 0$. It then turns out that the Hessian $\nabla^2 H$ of H as a linear map on the tangent space $T_{\zeta_0}(S^k)$ is evaluated as

$$\nabla^2 H(\eta) = \mathcal{A}^{-1}(\eta) - \mu_0 \eta, \quad \langle \eta, \zeta_0 \rangle = 0.$$

By using the present Hessian, we can show that the equilibrium ζ_0 is linearly stable, if ζ_0 is an eigenvector associated with the minimum or the maximum eigenvalue. For simplicity, we assume that all eigenvalues of $\mathcal{A}^{-1}$ are distinct. Since $\nabla H(\zeta_0) = 0$, the Hamiltonian is approximated, in the neighborhood of ζ_0, as

$$H(\zeta) = H(\zeta_0) + \frac{1}{2}\varepsilon^2 \langle \nabla^2 H(\eta), \eta \rangle = H(\zeta_0) + \frac{1}{2}\varepsilon^2 \langle \mathcal{A}^{-1}(\eta) - \mu_0 \eta, \eta \rangle.$$

Let ζ_0 be an eigenvector (matrix) associated with the minimum eigenvalue μ_0. Since H is a constant of motion, a solution $\zeta(t) = \zeta_0 + \varepsilon \eta(t)$ starting in the neighborhood of ζ_0 stays on a surface determined in $T_{\zeta_0}(S^k)$ by $H(\zeta) = H_0 + \varepsilon^2 h$ with $H_0 = H(\zeta_0)$ and $2h := \langle \nabla^2 H(\eta), \eta \rangle$. Incidentally, eigenvalues of $\nabla^2 H$ are $\mu - \mu_0$, where μ are eigenvalues of $\mathcal{A}^{-1}$. As μ_0 is the minimum eigenvalue by assumption, we have $\mu - \mu_0 \geq 0$. Hence, $\langle \mathcal{A}^{-1}(\eta) - \mu_0 \eta, \eta \rangle = 0$ if and only if η is parallel to ζ_0. However, η is a tangent vector to the sphere at ζ_0, the quadratic form $\langle \mathcal{A}^{-1}(\eta) - \mu_0 \eta, \eta \rangle$ is positive-definite on the tangent space $T_{\zeta_0}(S^k)$, so that the present quadratic surface $2h := \langle \nabla^2 H(\eta), \eta \rangle$ in the tangent space $T_{\zeta_0}(S^k)$ is elliptic. Therefore, solutions $\zeta(t) = \zeta_0 + \varepsilon \eta(t)$ near ζ_0 are bounded in the vicinity of ζ_0, which shows that the equilibrium ζ_0 is linearly stable. If ζ_0 is an eigenvector (matrix) associated with the maximum eigenvalue, a similar discussion runs in parallel to result in the same conclusion that ζ_0 is linearly stable. For $n = 3$, this fact is well known: Let $\mu_1 < \mu_2 < \mu_3$ be eigenvalues of the inertia tensor. Then, the equilibria associated with μ_1 and μ_3 are stable, but the equilibrium associated with μ_2 is unstable (see also Fig. 2.1) [2, 57].

Euler's equations are very general and highly generalized in connection with Lie algebras. Rigid bodies in n dimensions have been treated in [42, 70, 71] and the generalized Euler equations are studied in [18, 72].

5.8 Kaluza–Klein Formalism

Here we give a review of the Kaluza–Klein formalism [57] for the electron motion in a magnetic field $\boldsymbol{B}$, and further, for the sake of comparison, study the equations of motion for the planar three-body system.

Electron motion in a magnetic field
In the Kaluza-Klein formalism for electron motion, one introduces an auxiliary space $S^1 \cong SO(2)$ and extends the configuration space $\mathbb{R}^3$ to $\mathbb{R}^3 \times S^1$, which is a principal fiber bundle over $\mathbb{R}^3$ with the structure group $SO(2)$. Let (x_k, θ) be local

coordinates of $\mathbb{R}^3 \times S^1$. We define the connection form ω on $\mathbb{R}^3 \times S^1$ to be

$$\omega = \sum_{k=1}^{3} A_k dx_k + d\theta,$$

where $\boldsymbol{A} = (A_k)$ is a vector potential of the magnetic flux $\boldsymbol{B} = \nabla \times \boldsymbol{A}$. The vertical and the horizontal vector fields with respect to this connection are given by

$$\frac{\partial}{\partial\theta}, \quad X_k = \frac{\partial}{\partial x_k} - A_k \frac{\partial}{\partial\theta}, \quad k = 1, 2, 3,$$

respectively. The dual one-forms are ω and dx_k, satisfying

$$\omega(X_k) = 0, \quad \omega\Big(\frac{\partial}{\partial\theta}\Big) = 1, \quad dx_k(X_j) = \delta_{kj}, \quad dx_k\Big(\frac{\partial}{\partial\theta}\Big) = 0.$$

A canonical metric on $\mathbb{R}^3 \times S^1$ is defined to be

$$ds^2 = m \sum dx_k^2 + \omega^2,$$

where m is the mass of the electron. With respect to this metric, the vector fields X_k and $\partial/\partial\theta$ are mutually orthogonal.

While the standard local coordinates of $T(\mathbb{R}^3 \times S^1)$ are $(x_k, \theta, \dot{x}_k, \dot{\theta})$, we now introduce the variable

$$\pi = \sum A_k \dot{x}_k + \dot{\theta},$$

and take $(x_k, \theta, \dot{x}_k, \pi)$ as local coordinates of $T(\mathbb{R}^3 \times S^1)$. Then, in correspondence to the metric, the Lagrangian is expressed as

$$L = \frac{1}{2} m \sum \dot{x}_k^2 + \frac{1}{2}\pi^2.$$

We are in a position to take Hamel's approach to the equations of motion. Since $d(dx_k) = 0$ and

$$d\omega = \sum_{j<i} \Big(\frac{\partial A_i}{\partial x_j} - \frac{\partial A_j}{\partial x_i}\Big) dx_j \wedge dx_i,$$

the non-vanishing Hamel's symbols are

$$\gamma_{ij}^4 = \frac{\partial A_i}{\partial x_j} - \frac{\partial A_j}{\partial x_i} = \sum \varepsilon_{ikj} B_k, \quad i, j, k = 1, 2, 3,$$

and the other $\gamma^k_{\ell m}$ all vanish, where $k = 1, 2, 3$, and $\ell, m = 1, 2, 3, 4$. Since the Lagrangian is independent of x_k and θ, Hamel's equations of motion are expressed as

$$\frac{d}{dt}(m\dot{x}_k) + \pi \sum \varepsilon_{kij} B_i \dot{x}_j = 0, \quad \frac{d}{dt}\pi = 0.$$

This implies that π is a constant quantity, which we denote by q, a charge. Then, the first equation of the above is the same as Eq. (3.2) with $\boldsymbol{E} = 0$.

Equations of motion for the planar three-body system
We turn to the Lagrangian mechanics for a planar three-body system. The configuration space for the planar three-body system is a principal fiber bundle, $\dot{\mathbb{R}}^4 \to \dot{\mathbb{R}}^3$, with the fiber S^1 (see (1.57)), which means that the space S^1 is not auxiliary, in contrast to the Kaluza–Klein formalism stated above for the electron motion. We denote by x_k the Cartesian coordinates of $\dot{\mathbb{R}}^3$ in place of ξ_k (see (1.58)), and by θ an angular coordinate of S^1. Then, from (5.31), the connection form and the fundamental vector field F are expressed as

$$\omega = d\theta + \sum_{k=1}^{3} A_k dx_k, \quad F = \frac{\partial}{\partial \theta},$$

respectively, where A_k are the components of one of the vector potentials $\boldsymbol{A}_\pm$ given in (1.77) and where the θ used here is different form that used in (1.76). Vibrational vectors V_k are then given by

$$V_k = \frac{\partial}{\partial x_k} - A_k \frac{\partial}{\partial \theta}, \quad k = 1, 2, 3.$$

Then, the one-forms dx_k, ω and the vectors V_k, F are dual systems such that $dx_k(V_j) = \delta_{kj}, dx_k(F) = \omega(V_k) = 0, \omega(F) = 1$.

On introducing the variable π by

$$\pi = \dot{\theta} + \sum A_k \dot{x}_k,$$

we take $(x_k, \theta, \dot{x}_k, \pi)$ as local coordinates of the tangent bundle $T(\dot{\mathbb{R}}^4)$. Then, from (1.78), the Lagrangian L is defined to be

$$L = \frac{1}{8\rho} \sum \dot{x}_k^2 + \frac{1}{2}\rho\pi^2, \quad \rho = \sqrt{\sum x_k^2}.$$

In order to describe the equations of motion of Hamel's form, we need Hamel's symbols. The non-vanishing symbols are

$$\gamma^4_{kj} = F_{kj} = \sum \varepsilon_{k\ell j} B_\ell, \quad F_{kj} = \frac{\partial A_k}{\partial x_j} - \frac{\partial A_j}{\partial x_k}, \quad j,k,\ell \in \{1,2,3\},$$

and the other components all vanish, $\gamma^k_{\ell m} = 0$, $k = 1,2,3$, and $\ell, m \in \{1,2,3,4\}$, where $\boldsymbol{B} = (B_\ell)$ denotes the monopole field. It then turns out that the equations of motion are given by

$$\frac{d}{dt}\Big(\frac{1}{4\rho}\dot{x}_k\Big) + \sum F_{kj}\rho\pi\dot{x}_j - V_k(L) = 0,$$
$$\frac{d}{dt}(\rho\pi) = 0,$$

where use has been made of the fact that L is $SO(2)$-invariant; $F(L) = 0$. The last of the above equations means that $\rho\pi$ is a constant of motion, which we denote by q. Accordingly, the first equation above is rewritten as

$$\frac{d}{dt}\Big(\frac{1}{4\rho}\dot{\boldsymbol{x}}\Big) + q\boldsymbol{B}\times\dot{\boldsymbol{x}} - \frac{\partial L}{\partial \boldsymbol{x}} = 0, \quad \boldsymbol{x} = (x_k), \quad \boldsymbol{B} = \frac{1}{2}\frac{\boldsymbol{x}}{\rho^3}, \quad \rho = |\boldsymbol{x}|.$$

Further, it is straightforward to verify that the total energy E and the angular momentum $\boldsymbol{\Lambda}$ are conserved,

$$E = \frac{1}{8\rho}|\dot{\boldsymbol{x}}|^2 + \frac{q^2}{2\rho}, \quad \boldsymbol{\Lambda} = \boldsymbol{x}\times\frac{1}{4\rho}\dot{\boldsymbol{x}} - \frac{q}{2}\frac{\boldsymbol{x}}{\rho}.$$

The first and the second terms of the right-hand side of the first equation above are the kinetic energy associated with the metric (1.79) and a potential for repulsive force, respectively. From the second equation, we obtain $\boldsymbol{\Lambda}\cdot\boldsymbol{x}/\rho = -\frac{1}{2}q$, which means that the trajectories are sitting on the cone with the axis $\boldsymbol{\Lambda}$. See in addition [29] for the reduction of the Hamiltonian system on $T^*(\dot{\mathbb{R}}^4)$ by the $SO(2)$ symmetry. Another formulation of equations of motion for the planar three-body system is found in [61]. For the corresponding quantum system, the conserved quantity is quantized to take integer values [26].

5.9 Symplectic Approach to Hamilton's Equations

We derive Hamilton's equations of motion (4.61) in symplectic formalism. Let Θ denote the canonical one-form on $T^*(X_0)$. Following [62] and [31], we introduce the momentum variables p_i and ϖ_a by

$$p_i = \Theta\Big(\Big(\frac{\partial}{\partial q^i}\Big)^*\Big), \quad \varpi_a = \Theta(K_a), \tag{5.36}$$

respectively, where $(\partial/\partial q^i)^*$ are given in (4.35) and K_a are left-invariant vector fields on $SO(3)$. Both $(\partial/\partial q^i)^*$ and K_a can be viewed as locally defined on $T^*(X_0)$. Then, Θ is expressed as

$$\Theta = \sum_i p_i dq^i + \sum_a \varpi_a \omega^a. \tag{5.37}$$

The Hamiltonian vector field X_H is determined as usual through

$$\iota(X_H)d\Theta = -dH,$$

where ι denotes the interior product. We start by calculating $d\Theta$ to obtain

$$d\Theta = \sum dp_i \wedge dq^i + \sum d\varpi_a \wedge \omega^a + \sum \varpi_a d\omega^a,$$

and further calculate $d\omega^a$ with (4.34b) to find that

$$d\omega^a = -\sum_{b<c} \varepsilon_{abc} \Psi^b \wedge \Psi^c + \sum_{b<c} \varepsilon_{abc} \sum_i \Lambda_i^b dq^i \wedge \sum_j \Lambda_j^c dq^j + \sum_{i<j} \kappa_{ij}^a dq^i \wedge dq^j.$$

We assume that the Hamiltonian vector field X_H is expressed as

$$X_H = \sum \xi_i \Big(\frac{\partial}{\partial q^i}\Big)^* + \sum \eta_i \frac{\partial}{\partial p_i} + \sum \zeta_a K_a + \sum \chi_a \frac{\partial}{\partial \varpi_a}.$$

A straightforward calculation provides

$$\begin{aligned}
\iota(X_H)d\Theta = &\sum \eta_i dq^i - \sum_i \xi_i dp_i + \sum \chi_a(\Psi^a + \sum \Lambda_i^a dq^i) \\
&- \sum \zeta_a d\varpi_a - \sum \varepsilon_{cba} \varpi_c \zeta_b \Psi^a \\
&+ \sum \varepsilon_{cba} \varpi_c \Lambda_i^b \xi_i \Psi^a + \sum \varepsilon_{abc} \varpi_a \Lambda_j^b \Lambda_i^c \xi_j dq^i - \sum \varpi_a \kappa_{ij}^a \xi_j dq^i.
\end{aligned}$$

On the other hand, dH is expanded as

$$dH = \sum \Big(\frac{\partial}{\partial q^i}\Big)^* H dq^i + \sum \frac{\partial H}{\partial p_i} dp_i + \sum K_a(H)\Psi^a + \sum \frac{\partial H}{\partial \varpi^a} d\varpi_a.$$

Since $-dH = \iota(X_H)d\Theta$, we obtain

$$\begin{aligned}
-\Big(\frac{\partial}{\partial q^i}\Big)^* H &= \eta_i + \sum \chi_a \Lambda_i^a + \sum \varepsilon_{abc} \varpi_a \Lambda_j^b \Lambda_i^c \xi_j - \sum \varpi_a \kappa_{ij}^a \xi_j, \\
\frac{\partial H}{\partial p_i} &= \xi_i,
\end{aligned}$$

$$-K_a(H) = \chi_a - \sum \varepsilon_{acb}\varpi_c\zeta_b + \sum \varepsilon_{acb}\Lambda_i^b\xi_i,$$
$$\frac{\partial H}{\partial \varpi_a} = \zeta_a.$$

Thus, we find that

$$\xi_i = \frac{\partial H}{\partial p_i}, \qquad \zeta_a = \frac{\partial H}{\partial \varpi_a},$$
$$\chi_a = -K_a(H) + \sum \varepsilon_{acb}\varpi_c\frac{\partial H}{\partial \varpi_b} - \sum \varepsilon_{acb}\varpi_c\Lambda_i^b\frac{\partial H}{\partial p_i},$$
$$\eta_i = -\left(\frac{\partial}{\partial q^i}\right)^* H + \sum K_a(H)\Lambda_i^a - \sum \varepsilon_{acb}\varpi_c\frac{\partial H}{\partial \varpi_b}\Lambda_i^a + \sum \varpi_a\kappa_{ij}^a\frac{\partial H}{\partial p_j}.$$

If H is $SO(3)$ invariant, then one has $K_a(H) = \frac{d}{dt}H(\cdots, ge^{tR(\boldsymbol{e}_a)}, \cdot)|_{t=0} = 0$. It then turns out that Hamilton's equations of motion, $d\tilde{p}/dt = X_H(\tilde{p}), \tilde{p} \in T^*(X_0)$, are put in the form

$$\frac{dq^i}{dt} = \frac{\partial H}{\partial p_i}, \qquad \pi^a = \frac{\partial H}{\partial \varpi_a}, \tag{5.38a}$$
$$\frac{d\boldsymbol{\varpi}}{dt} = \boldsymbol{\varpi} \times \frac{\partial H}{\partial \boldsymbol{\varpi}} - \sum(\boldsymbol{\varpi} \times \boldsymbol{\lambda}_i)\frac{\partial H}{\partial p_i}, \tag{5.38b}$$
$$\frac{dp_i}{dt} = -\frac{\partial H}{\partial q^i} - \boldsymbol{\lambda}_i \cdot \left(\boldsymbol{\varpi} \times \frac{\partial H}{\partial \boldsymbol{\varpi}}\right) + \boldsymbol{\varpi} \cdot \sum \boldsymbol{\kappa}_{ij}\frac{\partial H}{\partial p_j}, \tag{5.38c}$$

where we note that the tangent vectors assigned to the respective coefficients, ξ_i and ζ_a, of $(\partial/\partial q^i)^*$ and K_a are $dq^i(d/dt) = \dot{q}^i$ and $\omega^a(d/dt) = \pi^a$ on account of (4.38) (see also (4.51)).

It is to be noted that the procedure done above is generic and verbatim applicable in deriving Eq. (2.101) if $(\partial/\partial q^i)^*$ and $\boldsymbol{\varpi} = \sum \varpi_a \boldsymbol{e}_a$ are replaced with Y_i and $\boldsymbol{M}$ given in (1.193) and in (2.99), respectively.

5.10 Remarks on Related Topics

5.10.1 Quantum Many-Body Systems

While this book deals with classical mechanics for many-body systems, quantum mechanics for many-body systems has long been of deep interest in molecular physics. In the Born–Oppenheimer approximation, molecules can be treated as many-body systems consisting of atoms together with potentials resulting from electron motions [53]. Let us take a triatomic molecule (e.g. that of water) as an example. The shape of the molecule is determined by two bond lengths, r_1, r_2, and the valence angle θ. The Hamiltonian operator for this molecule should be described in terms of the shape coordinates (r_1, r_2, θ). In general, the configuration space $\dot{Q}$ for an N-body system is made into a fiber bundle, $\dot{Q} \to \dot{M} = \dot{Q}/SO(3)$, with structure group $SO(3)$. The Hamiltonian operator of the triatomic molecule should be described in terms of shape coordinates like (r_1, r_2, θ) or be defined on the shape space $\dot{M}$ (see also the remark stated in the last paragraph of Sec. 1.9). In order to form a suitable Hilbert space, we need a vector bundle associated with $\dot{Q} \to \dot{M}$, which is assigned by the quantum number ℓ for the total angular momentum operator, $\boldsymbol{L}^2 = \ell(\ell + 1)$ or by choosing a representation space for $SO(3)$. Then, the Hilbert space is the space of square integrable sections of this bundle. The kinetic term of the Hamiltonian operator is obtained from the usual Laplacian on $\dot{Q}$ under a suitable procedure, and described in terms of the covariant derivation operators associated with the connection on $\dot{Q} \to \dot{M}$ [27, 35, 79, 80]. Physically speaking, this procedure is an application of the conservation of the total angular momentum, which serves to delete the rotational degrees of freedom. The quantum mechanics of two jointed cylinders with a no-twist condition has been studied in [33].

5.10.2 Geometric Phases and Further Reading

In relation to the geometry of many-body systems, deformable bodies and geometric phases are worth mentioning. It was Wilczek [83] who first studied deformable bodies from the gauge-theoretical point of view. See also [38] for pseudo-rigid bodies. Geometric phases are realizations of the connection theory in classical mechanics and quantum mechanics [12, 74] and further developed in geometric mechanics [81]. A basic reference on the application of connection theory is a paper [56], in which an effective use of connection theory is made in various examples from classical mechanics. While the connection introduced in Proposition 1.7.1 is called the Guichardet connection, the connection defined in a similar manner is referred to as a mechanical connection in [56].

A key idea behind these studies consists in a fibered manifold $\pi : Q \to M$ endowed with a distribution, where the distribution means a family of k-planes, that is, a linear subbundle of the tangent bundle of the manifold Q. A good short reference on this idea is the Introduction of [63], though all the chapters that follow it are worth reading of course. Further, the fibered manifold $\pi : Q \to M$ finds an application to classical mechanical systems with active constraints [55], where the fibers $\pi^{-1}(p)$, $p \in M$, are regarded as the states of the active constraint.

As is mentioned above, there are a variety of physical applications of connection theory [12], among which monopole bundles and instanton bundles are worth mentioning (see also [3, 24]). In particular, the monopole bundle is closely related to the planar three-body system discussed in Sect. 1.4. Since the connection form is independent of r (see (1.72) and (1.76)), the bundle $\dot{\mathbb{R}}^4 \to \dot{\mathbb{R}}^3$ reduces to the

Hopf bundle $S^3 \to S^2$, which is one of the monopole bundles ($U(1)$-bundles over S^2) discussed in [12]. All the monopole bundles or equivalently all the complex line bundles over S^2 with structure group $U(1)$ are parametrized by $n \in \mathbb{Z}$ [77]. While the complex line bundles over S^2 look mathematically simple, if discrete subgroups of $SO(3)$ and their unitary representations are taken into account as symmetry groups, together with some parameters, for Hermitian matrices defined on S^2, then the Chern numbers of the complex line bundles associated with eigenvalues exhibit interesting variations against the parameters [40, 41].

5.10.3 Open Dynamical Systems and Developments

One of the origins of port-Hamiltonian systems is a family of circuits with several ports. In [7], the circuit differential equations for currents and voltages are formulated on graphs consisting of nodes and branches. For an n-port R-circuit, a boundary condition for currents and voltages at the ports is given in terms of differential two-forms [7].

Dirac structures are introduced along with port-Hamiltonian systems to provide a unified description of power-conserving interconnections [73] in terms of port variables called flows and efforts like currents and voltages, respectively. For many mechanical systems, the state space $\mathcal{X}$ is a manifold and the flows corresponding to energy-storage are the elements, $-\dot{x}$, of the tangent bundle $T\mathcal{X}$, whereas the efforts are elements of the cotangent bundle $T^*\mathcal{X}$. The power-conserving relation can be described in terms of the symmetric bilinear form on $T_x(\mathcal{X}) \oplus T_x^*(\mathcal{X})$. A constant Dirac structure $\mathcal{D}_x$ is a maximal dimensional linear subspace of $T_x(\mathcal{X}) \oplus T_x^*(\mathcal{X})$ such that $\mathcal{D}_x = \mathcal{D}_x^\perp$ with respect to the symmetric bilinear form. Thus, Dirac structures on $\mathcal{X}$ are defined to be subbundles of $T\mathcal{X} \oplus T^*\mathcal{X}$ [13, 14] and draw much attention from a mechanical point of view [44, 85], though the nomenclature "Dirac structure" originates from Dirac brackets [57].

Bibliography

1. R. Abraham, J.E. Marsden, *Foundations of Mechanics*, 2nd edn. (Benjamin/Cummings, Reading, 1978)
2. V.I. Arnold, *Mathematical Methods of Classical Mechanics* (Springer, New York, 1978)
3. J.E. Avron, L. Sadun, J. Segert, B. Simon, Chern numbers, quaternions, and Berry's phases in Fermi systems. Commun. Math. Phys. **124**, 595–627 (1989)
4. A.M. Bloch, N.E. Leonard, J.E. Marsden, Controlled Lagrangian and the stabilization of mechanical systems I: The first matching theorem. IEEE Trans. Auto. Cont. **45**, 2253–2270 (2000)
5. A.M. Bloch, D.E. Chang, N.E. Leonard, J.E. Marsden, Controlled Lagrangian and the stabilization of mechanical systems II: Potential shaping, IEEE Trans. Auto. Cont. **46**, 1556–1571 (2001)
6. A.M Bloch, N.E. Leonard, J.E. Marsden, Controlled Lagrangians and the stabilization of Euler-Poincaré mechanical systems. Int. J. Robust Nonlinear Control **11**, 191–214 (2001)
7. R.K. Brayton, J.K. Moser, A theory of nonlinear networks - I; II. Quart. Appl. Math. **22**, 1–33; 81–104 (1964)
8. R.W. Brockett, L. Dai, Non-holonomic kinematics and the role of elliptic functions in constructive controllability, in *Nonholonomic Motion Planning*, ed. by Z. Li, J.F. Canny (Kluwer Academic, New York, 1993), pp 1–21
9. H. Cendra, D.D. Holm, J.E. Marsden, T.S. Ratiu, Lagrangian reduction, the Euler-Poincaré equations, and semidirect products. Am. Math. Soc. Transl. **186**, 1–25 (1998)
10. H. Cendra, J.E. Marsden, T. Ratiu, Lagrangian reduction by stages. Memoirs A.M.S. **152**(722) (2001)
11. W-L. Chow, Über Systeme von linearen partiellen Differentialgleichungen erster Ordnumg. Math. Anallen **67**, 98–105 (1939)
12. D. Chruściński, A. Jamiołkowski, *Geometric Phases in Classical and Quantum Mechanics* (Birkhäuser, Boston, 2004)
13. T. Courant, A. Weinstein, Beyond Poisson structures, in *Action hamiltoniennes de groupes. Troisiéme théorème de Lie (Lyon 1986), Travaux en Cours* ***27*** (Hermann, Paris, 1988), pp 39–49
14. T.J. Courant, Dirac manifolds. Trans. Am. Math. Soc. **319**, 631–661 (1990)
15. W.D. Curtis, F.R. Miller, *Differential Manifolds and Theoretical Physics* (Academic Press, Orlando, 1985)
16. R.H. Cushman, L.M. Bates, *Global Aspects of Classical Integrable Systems* (Birkhäuser, Basel, 1997)

T. Iwai, *Geometry, Mechanics, and Control in Action for the Falling Cat*,
Lecture Notes in Mathematics 2289, https://doi.org/10.1007/978-981-16-0688-5

17. C. Eckart, Some studies concerning rotating axes and polyatomic molecules. Phys. Rev. **47**, 552–558 (1935)
18. A.T. Fomenko, *Symplectic Geometry* (Gordon and Breach Science Publishers, New York, 1988)
19. K. Fujimoto, T. Sugie, Canonical transformation and stabilization of generalized Hamiltonian systems. Sys. Cont. Lett. **42**, 217–227 (2001)
20. X-S. Ge, Q-Z. Zhang, Optimal control of nonholonomic motion planning for a free-fall cat, in *Proceedings of the first International Conference on Innovative Computing*, Information and Control (ICICIC '06), 2006
21. H. Goldstein, *Classical Mechanics*, 2nd edn. (Addison-Wesley, Reading, 1980)
22. A. Guichardet, On rotation and vibration motions of molecules. Ann. Inst. H. Poincaré Phys. Théor. **40**, 329–342 (1984)
23. G. Hamel, *Theoretische Mechanik* (Springer, Berlin/Heidelberg, 1949)
24. Y. Hatsugai, Symmetry-protected $\mathbb{Z}_2$-quantization and quaternionic Berry connection with Kramers degeneracy. New J. Phys. **12**(21), 065004 (2010)
25. A. Hernández-Garduño, J.K. Lawson, J.E. Mardsen, Relative equilibria for the generalized rigid body. J. Geom. Phys. **53**, 259–274 (2005)
26. T. Iwai, A gauge theory for the quantum planar three-body problem. J. Math. Phys. **28**, 964–974 (1987)
27. T. Iwai, A geometric setting for internal motion of the quantum three-body system. J. Math. Phys. **28**, 1315–1326 (1987)
28. T. Iwai, A geometric setting for classical molecular dynamics. Ann. Inst. Henri Poincaré **47**, 199–219 (1987)
29. T. Iwai, N. Katayama, Two classes of dynamical systems all of whose bounded trajectories are closed. J. Math. Phys. **35**, 2914–2933 (1994)
30. T. Iwai, E. Watanabe, The Berry phase in the plate-ball problem. Phys. Lett. A **225**, 183–187 (1997)
31. T. Iwai, The mechanics and control for multi-particle systems. J. Phys. A **31**, 3849–3865 (1998)
32. T. Iwai, A. Tachibana, The geometry and mechanics of multi-particle systems. Ann. Inst. Henri Poincaré **70**, 525–559 (1999)
33. T. Iwai, Classical and quantum mechanics of jointed rigid bodies with vanishing total angular momentum. J. Math. Phys. **40**, 2381–2399 (1999)
34. T. Iwai, Geometric mechanics of many-body systems. J. Comput. Appl. Math. **140**, 403–422 (2002)
35. T. Iwai, H. Yamaoka, Stratified reduction of many-body kinetic energy operators. J. Math. Phys. **44**, 4411–4435 (2003)
36. T. Iwai, H. Yamaoka, Stratified reduction of classical many-body systems with symmetry. J. Phys. A Math. Gen. **38**, 2415–2439 (2005)
37. T. Iwai, H. Yamaoka, Stratified dynamical systems and their boundary behaviour for three bodies in space, with insight into vibrations. J. Phys. A Math. Gen. **38**, 5709–5730 (2005)
38. T. Iwai, The geometry and mechanics of generalized pseudo-rigid bodies. J. Phys. A Math. Theor. **43**(28), 095206 (2010)
39. T. Iwai, H. Matsunaka, The falling cat as a port-controlled Hamiltonian system. J. Geom. Phys. **62**, 279–291 (2012)
40. T. Iwai, B. Zhilinskii, Energy bands: Chern numbers and symmetry. Ann. Phys. **326**, 3013–3066 (2011)
41. T. Iwai, B. Zhilinskii, Chern number modification in crossing the boundary between different band structures: Three-band models with cubic symmetry. Rev. Math. Phys. **29**(91), 1750004 (2017)
42. A. Izosimov, Stability of relative equilibria of multidimensional rigid body. Nonlinearity **27**, 1419–1443 (2014)
43. J.V. José, E.J. Saletan, *Classical Dynamics, A Contemporary Approach* (Cambridge University Press, Cambridge, 1998)

44. M. Jotz, T.S. Ratiu, Dirac structures, nonholonomic systems and reduction. Rep. Math. Phys. **69**, 5–56 (2012)
45. V. Jurdjevic, The geometry of plate-ball problem. Arch. Rational Mech. Anal. **124**, 305–328 (1993)
46. T.R. Kane, M.P. Scher, A dynamical explanation of the falling cat phenomena. Int. J. Solid Struct. **5**, 663–670 (1969)
47. D.G. Kendall, A survey of the statistical theory of shape. Stat. Sci. **4**, 87–120 (1989)
48. S. Kobayashi, K. Nomizu, *Foundations of Differential Geometry*, Vol. I (Interscience Publication, New York, 1963)
49. J. Koiller, Reduction of some classical non-holonomic systems with symmetry. Arch. Rat. Mech. Anal. **118**, 113–148 (1992)
50. W.-S. Koon, J.E. Marsden, Optimal control for holonomic and nonholonomic mechanical system with symmetry and Lagrangian reduction. SIAM J. Control Optim. **35**, 901–929 (1997)
51. J-M. Lévy-Leblond, M. Lévy-Nahas, Three-particle nonrelativistic kinematics and phase space. J. Math. Phys. **6**, 1571–1575 (1965)
52. R.G. Littlejohn, M. Reinsch, Internal or shape coordinates in the n-body problem. Phys. Rev. **A 52**, 2035–2051 (1995)
53. R.G. Littlejohn, M. Reinsch, Gauge fields in the separation of rotations and internal motions in the n-body problem. Rev. Mod. Phys. **69**, 213–275 (1997)
54. R.G. Littlejohn, K.A. Mitchell, Gauge theory of small vibrations of polyatomic molecules, in *Geometry, Mechanics, and Dynamics*, ed. by P. Newton, P. Holmes, A. Weinstein (Springer, New York, 2002)
55. C-M. Marle, Géométrie des systèmes mécaniques à liaisons actives, in *Symplectic Geometry and Mathematical Physics*, ed. by P. Donato, C. Duval, J. Elhadad, G.M. Tuynman (Birkhäuser, Boston, 1991)
56. J.E. Marsden, R. Montgomery, T. Ratiu, *Reduction, Symmetry, and Phases in Mechanics*, vol. 88, no. 436 (Memoirs of AMS, Providence, 1990)
57. J.E. Marsden, T.S. Ratiu, *Introduction to Mechanics and Symmetry* (Springer, New-York, 1994)
58. J.E. Marsden, M. Perlmutter, The orbit bundle picture of cotangent bundle reduction. C. R. Math. Rep. Acad. Sci. Can. **22**, 33–54 (2000)
59. K.A. Mitchell, R.G. Littlejohn, The rovibrational kinetic energy for complexes of rigid molecules. Mol. Phys. **96**, 1305–1315 (1999)
60. R. Montgomery, Canonical formulations of a classical particle in a Yang-Mills field and Wong's equations. Lett. Math. Phys. **8**, 59–67 (1984)
61. R. Montgomery, Optimal control of deformable bodies and its relation to gauge theory. *The Geometry of Hamiltonian Systems*, ed. by T. Ratiu (Springer, New York, 1991)
62. R. Montgomery, Gauge theory of the falling cat, in *Dynamics and Control of Mechanical Systems*, ed. by M.J. Enos, (American Mathematical Society, Providence, 1993), pp 193–218
63. R. Montgomery, *A Tour of Subriemannian Geometries, Their Geodesics and Applications* (American Mathematical Society, A Tour of Subriemannian Geometries, Their Geodesics and Applications, 2002)
64. M. Nakahara, *Geometry, Topology and Physics* (Adam Hilger, Bristol, 1990)
65. M.S. Narasimhan, T.R. Ramadas, Geometry of $SU(2)$ gauge fields. Commun. Math. Phys. **67**, 121–136 (1979)
66. R. Ortega, A.J. van der Schaft, I. Mareels, B. Maschke, Putting energy back in control. IEEE Control Syst. Mag. **21**, 18–33 (2001)
67. R. Ortega, A. van der Schaft, B. Maschke, G. Escobar, Interconnection and damping assignment passivity-based control of port-controlled Hamiltonian systems. Automatica **38**, 585–596 (2002)
68. J-P. Ortega, T.S. Ratiu, *Momentum Maps and Hamiltonian Reduction* (Birkhäuser, Boston, 2004)
69. J.W. Polderman, J.C. Willems, *Introduction to Mathematical Systems Theory* (Springer, New York, 1998)

70. T. Ratiu, The motion of the free n-dimensional rigid body. Indiana Univ. Math. J. **29**, 609–629 (1980)
71. T. Ratiu, Euler-Poisson equations on Lie algebras and the N-dimensional heavy rigid body. Am. J. Math. **104**, 409–448 (1982)
72. T. Ratiu, D. Tarama, The $U(n)$ free rigid body: Integrability and stability analysis of the equilibria. J. Diff. Eq. **259**, 7284–7331 (2015)
73. A. van der Schaft, Port-Hamiltonian systems: an introductory survey, in *Proceedings of the International Congress of Mathematicians, Madrid, Spain*, Vol. III (European Mathematical Society, Zurich, 2006)
74. A. Shapere, F. Wilczek (eds.), *Geometric Phases in Physics* (World Scientific, Singapore, 1989)
75. R.W. Sharpe, *Differential Geometry* (Springer, New York, 1997)
76. E. Straume, A geometric study of many body systems, Lobachevskii J. Math. **24**, 73–134 (2006)
77. N. Steenrod, *The Topology of Fiber Bundles* (Princeton University Press, Princeton, New Jersey, 1951)
78. H.J. Sussmann, V. Jurdjevic, Controllability of nonlinear systems. J. Diff. Eqs. **12**, 95–116 (1972)
79. A. Tachibana, T. Iwai, Complete molecular Hamiltonian based on the Born-Oppenheimer adiabatic approximation. Phys. Rev. **A33**, 2262–2269 (1986)
80. S. Tanimura, T. Iwai, Reduction of quantum systems on Riemannian manifolds with symmetry and application to molecular mechanics, J. Math. Phys. **41**, 1814–1842 (2000)
81. A. Weinstein, Connection of Berry and Hannay type for moving Lagrangian submanifolds. Adv. Math. **82**, 133–159 (1990)
82. E.T. Whittaker, *A Treatise on the Analytical Dynamics of Particles and Rigid Bodies* (Cambridge University Press, London, 1959)
83. F. Wilczek, Gauge theory of deformable bodies, in *XVIIth International Col. on Group Theoretical Methods in Physics*, ed. by Y. Saint-Aubin, L. Vinet (World Scientific, Singapore, 1989), pp. 154–167
84. S.K. Wong, Field and particle equations for the classical Yang-Mills field and particles with isotopic spin. Nuovo Cimento **65a**, 689–694 (1970)
85. H. Yoshimura, J.E. Marsden, Dirac structures in Lagrangian mechanics, Part I: Implicit Lagrangian systems; Part II: Variational structures, J. Geom. Phys. **57**, 133–156; 209–250 (2006)

Index

T. Iwai, *Geometry, Mechanics, and Control in Action for the Falling Cat*,
Lecture Notes in Mathematics 2289, https://doi.org/10.1007/978-981-16-0688-5

LECTURE NOTES IN MATHEMATICS

Springer

Editorial Policy

1. Lecture Notes aim to report new developments in all areas of mathematics and their applications – quickly, informally and at a high level. Mathematical texts analysing new developments in modelling and numerical simulation are welcome.

   Manuscripts should be reasonably self-contained and rounded off. Thus they may, and often will, present not only results of the author but also related work by other people. They may be based on specialised lecture courses. Furthermore, the manuscripts should provide sufficient motivation, examples and applications. This clearly distinguishes Lecture Notes from journal articles or technical reports which normally are very concise. Articles intended for a journal but too long to be accepted by most journals, usually do not have this "lecture notes" character. For similar reasons it is unusual for doctoral theses to be accepted for the Lecture Notes series, though habilitation theses may be appropriate.

2. Besides monographs, multi-author manuscripts resulting from SUMMER SCHOOLS or similar INTENSIVE COURSES are welcome, provided their objective was held to present an active mathematical topic to an audience at the beginning or intermediate graduate level (a list of participants should be provided).

   The resulting manuscript should not be just a collection of course notes, but should require advance planning and coordination among the main lecturers. The subject matter should dictate the structure of the book. This structure should be motivated and explained in a scientific introduction, and the notation, references, index and formulation of results should be, if possible, unified by the editors. Each contribution should have an abstract and an introduction referring to the other contributions. In other words, more preparatory work must go into a multi-authored volume than simply assembling a disparate collection of papers, communicated at the event.

3. Manuscripts should be submitted either online at www.editorialmanager.com/lnm to Springer's mathematics editorial in Heidelberg, or electronically to one of the series editors. Authors should be aware that incomplete or insufficiently close-to-final manuscripts almost always result in longer refereeing times and nevertheless unclear referees' recommendations, making further refereeing of a final draft necessary. The strict minimum amount of material that will be considered should include a detailed outline describing the planned contents of each chapter, a bibliography and several sample chapters. Parallel submission of a manuscript to another publisher while under consideration for LNM is not acceptable and can lead to rejection.

4. In general, **monographs** will be sent out to at least 2 external referees for evaluation.

   A final decision to publish can be made only on the basis of the complete manuscript, however a refereeing process leading to a preliminary decision can be based on a pre-final or incomplete manuscript.

   Volume Editors of **multi-author works** are expected to arrange for the refereeing, to the usual scientific standards, of the individual contributions. If the resulting reports can be forwarded to the LNM Editorial Board, this is very helpful. If no reports are forwarded or if other questions remain unclear in respect of homogeneity etc, the series editors may wish to consult external referees for an overall evaluation of the volume.

5. Manuscripts should in general be submitted in English. Final manuscripts should contain at least 100 pages of mathematical text and should always include

   – a table of contents;
   – an informative introduction, with adequate motivation and perhaps some historical remarks: it should be accessible to a reader not intimately familiar with the topic treated;
   – a subject index: as a rule this is genuinely helpful for the reader.
   – For evaluation purposes, manuscripts should be submitted as pdf files.

6. Careful preparation of the manuscripts will help keep production time short besides ensuring satisfactory appearance of the finished book in print and online. After acceptance of the manuscript authors will be asked to prepare the final LaTeX source files (see LaTeX templates online: https://www.springer.com/gb/authors-editors/book-authors-editors/manuscriptpreparation/5636) plus the corresponding pdf- or zipped ps-file. The LaTeX source files are essential for producing the full-text online version of the book, see http://link.springer.com/bookseries/304 for the existing online volumes of LNM). The technical production of a Lecture Notes volume takes approximately 12 weeks. Additional instructions, if necessary, are available on request from lnm@springer.com.

7. Authors receive a total of 30 free copies of their volume and free access to their book on SpringerLink, but no royalties. They are entitled to a discount of 33.3 % on the price of Springer books purchased for their personal use, if ordering directly from Springer.

8. Commitment to publish is made by a *Publishing Agreement*; contributing authors of multiauthor books are requested to sign a *Consent to Publish form*. Springer-Verlag registers the copyright for each volume. Authors are free to reuse material contained in their LNM volumes in later publications: a brief written (or e-mail) request for formal permission is sufficient.

Addresses:

Professor Jean-Michel Morel, CMLA, École Normale Supérieure de Cachan, France
E-mail: moreljeanmichel@gmail.com

Professor Bernard Teissier, Equipe Géométrie et Dynamique,
Institut de Mathématiques de Jussieu – Paris Rive Gauche, Paris, France
E-mail: bernard.teissier@imj-prg.fr

Springer: Ute McCrory, Mathematics, Heidelberg, Germany,
E-mail: lnm@springer.com

Printed in the United States
by Baker & Taylor Publisher Services